Surviving the ASIC Experience

John Schroeter

PRENTICE HALL, Englewood Cliffs, New Jersey 07632

Library of Congress Cataloging-in-Publication Data

Schroeter, John.
    Surviving the ASIC experience / John Schroeter.
        p.   cm.
    Includes index.
    ISBN 0-13-877838-8
    1. Integrated circuits industry. 2. Application specific
integrated circuits.   I. Title.
HD9696.I582S37   1992
338.4'762138815--dc20   91-24437
                                                            CIP

Editorial/production supervision: MARY P. ROTTINO
Cover design: BUTLER/UDELL
Manufacturing buyer: SUSAN BRUNKE
Pre-press buyer: MARY E. MCCARTNEY
Acquisitions editor: KAREN GETTMAN

© 1992 by Prentice-Hall, Inc.
A Simon & Schuster Company
Englewood Cliffs, New Jersey 07632

The publisher offers discounts on this book when ordered
in bulk quantities. For more information, write:

    Special Sales/Professional Marketing
    Prentice-Hall, Inc.
    Professional & Technical Reference Division
    Englewood Cliffs, New Jersey 07632

Printed in the United States of America

10 9 8 7 6 5 4 3 2 1

ISBN 0-13-877838-8

Prentice-Hall International (UK) Limited, London
Prentice-Hall of Australia Pty. Limited, Sydney
Prentice-Hall Canada Inc., Toronto
Prentice-Hall Hispanoamericana, S.A., Mexico
Prentice-Hall of India Private Limited, New Delhi
Prentice-Hall of Japan, Inc., Tokyo
Simon & Schuster Asia Pte. Ltd., Singapore
Editora Prentice-Hall do Brasil, Ltda., Rio de Janeiro

To Jolene, Jenna, and JoBeth

Contents

Preface   ix

1   Introduction to ASIC Issues   1

*The ASIC Industry—ASICs-Why, Why Not?—ASIC
Advantages—ASIC Disadvantages—Overcoming the
Barriers*

2   Selecting the ASIC Methodology   7

*The ASIC Design Flow—ASIC Design Methodology
Choices and Trade-offs—Custom—Full-Custom
Cell-based—Array-based—Gate Array Architectures
Counting the Gates—Estimating Performance—Integrating
RAM—Programmable Logic—Programmable Logic Is
Low Risk—FPGA Architectures—Evaluating PLD
Performance—Cost Effectiveness—FPGA Development
Tools—Design Validation Using FPGAs*

3   Selecting the ASIC Technology   34

*CMOS—Process Design Rule Considerations—Step
Coverage—Performance—BiCMOS—Load-Driving
Capability—Ground Bounce—Cost Impact of BiCMOS
Processing—Bipolar—GaAs—GaAs Technology
GaAs on Silicon—Radiation-hardened Processes—EMP
Effects—Ionizing Dose Rate—Total Dose—Neutron
Fluence—Single-event Upset—Design for Radiation
Environments—Rad Hard or Rad Tolerant?*

4 **Selecting the ASIC Package**   48

*Package Trade-offs—Ceramic versus Plastic
Surface-mounted Packages—Fine Lead Pitch
Considerations—Chip Carriers—Through-hole
Mounted Packages—Dual-inline Packages—Pin
Grid Arrays—Tape-automated Bonding—TAB
Manufacturing Sequence—Chip on Board—Multichip
Packages—Electrical Considerations—Thermal
Considerations*

5 **Selecting the ASIC Design Tools**   67

*Schematic Capture—Logic Synthesis—Evaluating Logic
Synthesis Tools—Modeling Behavior—VHDL—Simulation
Simulation Overview—Simulation Vector—Generation
ASIC Cell Library Considerations—Debugging the Design
Simulating Designs in Submicron Processes—Timing
Analysis—Hardware Modeling—ASIC Emulation—Physical
Design—Floor Planning—Layout Optimization—The ASIC
Design Environment—Maintenance and Upgrades—Design
Tool Integration—EDIF—Frameworks—Workstation
Hardware—Design Tool Quality*

6 **Selecting the ASIC Vendor**   98

*Preparing the Request for Proposal—Evaluating the Bidders
Technical Issues—Cost Issues—Design and Prototyping
Design Tools—Production Issues—Business Issues
Contracting with ASIC Suppliers*

7 **Design Guidelines and Issues**   111

*Partitioning—Including Core Functions on the ASIC
Mixed-signal Design—Mixed-signal Simulation and Test Issues
Mixed-signal Process Considerations—The ASIC Specification
Design Guidelines—I/O Considerations—Simulation
Considerations—Basic Design Considerations—Design for
Reliability—Design for Testability—Tester Considerations
Improving Test Coverage—Fault Grading—Scan Test
JTAG Boundary Scan—Automatic Test Pattern Generation
Test Generation for Memory—Testability Trade-offs—Project
Management—The Project Schedule—Design Reviews
Time-to-Market Considerations*

8 ASIC Prototyping and Verification 142

*Design Verification—Prototype Fabrication—Mask
Making—Wafer Fabrication—Wafer Probing—Packaging
Prototype Test—Test Program Development—Tester Basics
At-speed Testing—Debugging the Test Setup—Characterization
Evaluating ASIC Testers*

9 ASIC Production Issues 155

*The Preproduction Phase—Production Planning—Quality
Assurance—Military Standard Quality and Reliability
Statistical Process Control—Electrostatic Discharge Controls
Failure Analysis—Fab Capacity and Product Allocation
Alternate Sourcing*

10 ASIC Cost Determination 168

*ASIC Production Unit Cost Factors—Estimating the Die Size
Predicting Process Yield—Calculating the Production Unit Price*

Appendix A 180

ASIC Design Evaluation and Pricing Programs

Appendix B 181

Directory of ASIC Vendors

Appendix C 185

Directory of Programmable Logic Vendors

Appendix D 186

Directory of CAE Tool Vendors

Appendix E 189

Directory of Package and Assembly Vendors

Appendix F 190

Glossary of ASIC Terms and Acronyms

Appendix G 204

Read More About It

Preface

IF INDUSTRY ANALYSTS ARE CORRECT, application-specific integrated circuits (ASICs) will be the primary medium for the design of electronic products by the end of this century. As evidenced by a burgeoning ASIC industry and a plethora of trade publications dedicated to covering the latest and greatest in ASIC technology offerings, one could quickly surmise that we are in the midst of an ASIC revolution.

There are many reasons to consider ASICs. In fact, many applications require the use of ASICs to realize basic product objectives. In many cases, the use of ASICs allows system designs to be tailored to specific markets with features that are simply unavailable in standard ICs. Other benefits afforded by ASICs include improvements in performance, reliability, power, system size, cost, and market competitiveness. In addition, ASIC offerings cover the gamut from applications requiring the integration of only a few hundred gates of random logic to those that play host to entire subsystems, consisting of 100,000 gates or more. Such ASICs may integrate entire processors, memory, state machines, and other complex, high performance functions. For these and other reasons, ASICs have long been predicted to replace standard SSI and MSI components.

Yet, with all this going for it, fewer than 10% of all U.S. electronics engineers have ever been involved with ASICs to any degree. Pardner Wynn, of Xylinx, observes "The much heralded ASIC revolution is still a faint noise." Why then hasn't ASIC technology been embraced by the majority of candidate engineers who could stand to benefit so much? Oddly, one reason is that the technology itself is advancing at such a rapid pace. The choices of products and technologies as well as the number of variables that must be evaluated are expanding at a seemingly exponential rate. Furthermore, a tremendous effort is required to understand the tools of ASIC implementation. A great amount of time is needed to research, acquire, install, train, manage, and successfully use the new tools, as well as to understand and deal with the entirely new set of problems they present.

Another significant factor impacting acceptance is the perceived risk that ASICs impose. Today, there is a 50 - 50 chance that an ASIC will be fully operational when first introduced into the target system. With the high development costs and long leadtimes of ASICs, it is no wonder that engineering managers are reluctant to risk delaying and jeopardizing their design projects on what is perceived to be a relatively unproved technology. As a result, many engineers are locked into traditional design methodologies and the ASIC revolution continues to pass them by.

This phenomenon has another interesting effect which can be seen in the way ASIC design tools are developing. CAE vendors tend to cater to the needs of the experienced ASIC designers. This only makes sense since they are the ones buying and using the tools. The more experience these users gain, the more sophisticated they become. Ultimately, it is these users who drive the development of new CAE tools. In this sense, the tools are becoming more mature, but only from the perspective of these "power users," not of the mainstream engineers who are just getting their feet wet in ASIC technology. As this disparity increases, adoption becomes even more difficult.

However difficult it may be, adoption is inevitable. As newer generations of electronic products are introduced that take advantage of ASIC benefits, competing companies will be forced to comply if they are to survive, let alone remain competitive.

To survive the competition, though, we must first survive the initial ASIC experience. To this end, this book is a bridge. In it, we will examine the barriers to adopting ASIC technology. We will offer insightful solutions to these problems, as well as provide the tools and information needed to sort through and understand the myriad issues that will inevitably be faced when entering the ASIC arena. Each chapter deals with its subject matter from the perspective of how to effectively evaluate, qualify, negotiate, and manage each step in the ASIC design process to its successful completion.

The introductory chapter, *Introduction to ASIC Issues*, provides a baseline perspective on the ASIC industry and explores the pros and cons of the ASIC approach to electronic system design. The next five chapters deal with the critical decisions that will have to be made before the ASIC design project can even begin. These include selection of the optimum design methodology (Chapter 2), process technology (Chapter 3), device package (Chapter 4), design tools (Chapter 5), and ASIC vendor (Chapter 6). Chapter 7, *Design Guidelines and Issues*, is a particularly important chapter in that it sheds light on the most common errors committed in ASIC design (made by experienced and novice designers alike), which complicate and jeopardize the majority of ASIC design projects. Chapter 8, *ASIC Prototyping and Verification*, presents strategies for optimizing the prototyping, design verification, and debugging processes. The issues and strategies involved in successful ASIC procurement are presented in Chapter 9, *ASIC Production Issues*. Because ASICs are typically sole-sourced devices, their procurement is not as straightforward as that for multisourced, standard catalog components. Successful ASIC procurement requires an understanding of the issues, as well as special planning and tactics if a continuous and uninterrupted supply of parts is to be secured. Many a product (and company) has met its demise due to the inability of the ASIC supplier to deliver working ASICs on time. Finally, Chapter 10, *ASIC Cost Determination*, provides the necessary information for determining the fair market price for the desired ASIC configuration. The reader will learn to estimate an ASIC design's corresponding die size, predict the process yield and generate quotes for comparison and calibration with the actual quotes that will be received from candidate ASIC vendors bidding on the project. The material in this chapter will also enable the reader to evaluate different partitioning schemes for an ASIC design, perform trade-off and budget analysis, identify potential yield and delivery problems, and negotiate costs from a position of strength. The PC-based programs described in Appendix A, *ASIC Design Evaluation and Pricing Programs*, provide an excellent companion to this and other chapters in *Surviving the ASIC Experience*. Additional appendix provide directories of ASIC, programmable logic, CAE tool, and package vendors. A glossary listing important ASIC terms and definitions is also provided.

Equally as important as knowing how to win with ASICs is knowing how not to lose. The material in this book will lead to an understanding of the problems, the reasons for the failures, and how the many potential pitfalls can be avoided.

If properly executed, the strategies presented in *Surviving the ASIC Experience* will tremendously increase the ASIC user's chances of success. Achieving *complete* ASIC success is our ultimate goal—success as measured by the chip working on first pass, at speed, in the system (not only on a test fixture), on time, within budget, and without pain. It's that elusive state of being that Mehendra Jain, of <u>ASIC Technology & News,</u> refers to as ASIC nirvana.

Chapter 1

Introduction to ASIC Issues

BY NOW, EVERYONE INVOLVED IN ELECTRONICS is familiar with ASICs. But few can agree on a precise definition of the term. The application specific integrated circuit label has been applied to virtually every type of chip that was designed to perform a dedicated task. Depending on one's perspective, ASICs can be ordered from a catalog, shipped from stock, programmed at the bench, used by a single customer or by many customers. Their functions may be defined by the user or by the supplier.

We can narrow our definition, though, if we view ASICs according to their intended use. For example, many merchant market suppliers offer standard catalog components based on custom and semicustom design methodologies. These chips usually address a specific application in niche markets. This class of ASICs might be more accurately described as application specific standard ICs. Conversely, ASICs designed by the end-user, *specifically for his proprietary application*, represent the class of ASICs with which we will be concerned. In our definition, the term applies regardless of the technology or methodology employed.

The ASIC Industry

The relatively short history of ASICs has indeed been volatile. Company mergers and exits, new products, processes, and methodologies are announced almost daily. And because the ASIC industry moves at such a frenzied pace, it receives a disproportionate share of trade media coverage. As a result, it is also subject to a tremendous volume of hype. New technologies are announced and reported long in advance of their general availability—an occurrence that does more to serve the positioning strategies of the vendors than it does for the targeted customers.

Forecasts for the industry have been equally volatile. The ASIC market is a moving target and therefore, is difficult to pin down. As a result, most of the predictions made by industry analysts haven't panned out. For example, during the mid-1980s, it was anticipated that ASICs would capture 50% of the total semiconductor market by 1990. When 1990 rolled around, the number was closer to 10%. Another forecast stated that all-layer, cell-based methodologies would overtake array-based methodologies in the same forecast period. Cell-based ASICs still contribute only a small percentage of the revenue mix for the major ASIC suppliers.

Although most of the electronics media attention has been focused on 100,000-gate, high-performance ASIC offerings, only a small percentage of new ASIC design starts actually utilize them. The vast majority of ASIC designs are relatively simple and use mainstream, stable technologies. There will, however, always be those who push the limits. It is these relatively few, but important applications that drive the technology, as well as the EDA (electronic design automation) tools that are used to design them. Users in the high-complexity segment are typically implementing custom CPUs, complete with on-chip memory and peripheral control and interface logic, which are effectively high-performance systems on a chip. Mainstream users, on the other hand, have historically been content with logic consolidation—the majority of ASIC applications. In fact, two-thirds of all design starts consist of fewer than 20,000 gates, with a median gate count of about 15,000 gates. The general level of integration as a whole, though, is on the rise. (This trend seems to follow for the complexity of all types of ICs; as feature sizes shrink, the average density tends to rise).

Although adoption of the technology is off to a slower than anticipated start, the pace of new design starts is picking up. Also, the industry is beginning to mature and, perhaps, even settle down a bit as the roles of suppliers become better defined. This trend will continue as the industry responds to the realities of the marketplace.

2

Customers, though, will always be the driving force behind developments in the industry. In the ASIC world, its the diversity of ASIC users that has done the most to shape it. Although the market is dominated by a few large, broad-based semiconductor companies, the wide range of customer requirements has created niche opportunities for hundreds of smaller suppliers. But just as potential customers are key to market growth, they have also been the major bottleneck to a more rapidly expanding ASIC market. The vast majority of electronic design engineers have never designed an ASIC. The expectation that system designers would naturally (and quickly) become ASIC designers has gone largely unfulfilled.

Whatever the reasons for avoiding ASICs, they can generally be boiled down to unacceptable levels of cost and risk. Until these issues are adequately addressed, widespread acceptance of the technology will continue to be unnecessarily restricted. In fact, many designers don't want to use or increase their use of ASICs because they fear missed schedules and, ultimately, the failure of the design project. Cadence, in a recent advertisement, described the situation best when they wrote, "How you play the game determines if you win or lose. And there's plenty at stake. Your design. Your project. Maybe even your company." Some game! Unknown to most ASIC design candidates, though, is that nearly all the risk and cost factors affecting ASIC design can be controlled by the designer, provided he is equipped with the proper tools and knowledge.

ASICs—Why, Why Not?

The relative advantages and disadvantages of ASICs, as they apply to any one potential user, are determined by the nature of the proposed application. Factors representing incentives or barriers to adoption may include the product development budget, the available in-house design expertise, anticipated production volume, desired product features and, of course, the competition. A significant amount of analysis is often necessary to determine if the use of an ASIC is appropriate. The particular *mix* of advantages and disadvantages that is relevant will depend on the design goals. For example, for producers in the consumer or auto industry segments, cost reduction is perhaps the most significant benefit. Performance will be the critical advantage for computer makers. In military, aerospace, and medical markets, size and reliability may be the most important attributes. The following discussion offers considerations and trade-offs that, when carefully weighed in the context of the application, should yield an appropriate decision regarding ASIC adoption.

ASIC Advantages

• In many cases, ASICs represent the only way to implement a given design. The desired performance and functionality may simply be unattainable using standard components. In such cases, ASICs become the enabling technology.

• ASICs enable the inclusion of unique features that may add value and differentiation from competing products.

• A particular application may impose size constraints or space limitations. A single ASIC can replace a large number of standard components, thus integrating entire printed circuit boards. The use of ASICs may also allow a greater number of features in a smaller space.

• The reduction in the system component count can decrease system costs, increase reliability, and lower the power and cooling requirements. The lower power of some ASICs may also enable battery operation.

• ASICs ease inventory management by providing a lower number of unique parts.

• Taking the ASIC approach to system design may reduce the development time, thereby accelerating product introduction.

• ASICs may provide substantial increases in system performance and throughput over the equivalent implementation in standard components.

• ASICs greatly enhance system design security, thus making reverse engineering extremely difficult if not impossible.

ASIC Disadvantages

• The cost of prototyping, or nonrecurring engineering (NRE), can be quite high, depending on the complexity of the design and the method of implementation.

• Although time to market is an often cited ASIC advantage, the schedule risk imposed by the possibility of iteration can have a significant impact on product introduction plans. This is further exacerbated by shrinking product life-cycles and narrow market entry windows.

• It is an often reported statistic that nearly 50% of all ASIC designs fail to operate in the target system on first pass.

• It is difficult to fine-tune the design in the latter stages of development. There is also considerable difficulty in making design changes as the product evolves.

• Testing and design debug can be extremely difficult due to the inability to probe nodes that are internal to the ASIC.

• Multisourcing can be difficult and expensive. Being tied to a single supplier offers little leverage in purchasing.

• Production volumes may never reach the breakeven quantity, making the ASIC implementation less cost effective than a standard components solution.

• The inability to integrate desired functions may preclude the use of an ASIC.

Overcoming the Barriers

Although the list of disadvantages can be discouraging, many of these shortcomings are being resolved. New design methodologies offering faster, less expensive (and therefore less risky) prototyping are emerging. Larger wafer sizes and better yields are producing lower production costs. Design tools are maturing and becoming more reliable and easier to use.

While most design projects will enjoy many of the benefits ASICs offer, virtually every design team will have to contend with the problems they can impose. Minimizing the difficulties, though, can be accomplished through careful planning and management. Prerequisite is a complete understanding of the design process, the critical decisions and activities, as well as the many dangers and pitfalls that will inevitably be encountered along the way. Developing this understanding is the focus of the chapters that follow.

Chapter 2

Selecting the ASIC Methodology

EVEN MORE COMPLEX than determining whether to apply an ASIC solution is determining *which* ASIC solution to use. There are fundamental differences in the integration level, gate count, design tools, method of design verification, and development and production costs among the various ASIC approaches. These factors play a significant role in the selection of an ASIC methodology, technology, and vendor.

The selection criteria should be based on the proposed approach's ability to satisfy all design objectives, provide the shortest possible time to market, and do so at the lowest possible development and production costs. The mix of process technology, design methodology, and tools will also be a function of the application. For example, computer applications involving custom high-speed digital architectures are more likely to use submicron 100K gate arrays and high-level design tools. Industrial controllers are likely to integrate digital and analog functions on the same chip, using a cell-based design methodology in a less aggressive process that is optimized more for analog accuracy than for digital speed. Consumer products, with their high production volumes, are extremely cost sensitive and may call for a full-custom design approach in order to achieve the smallest possible die size.

In this chapter and the next, an overview of the dominant design methodology and process technology options will be provided. The relative strengths and weaknesses of each will be revealed when studied in the context of the target application. All have different characteristics and produce significantly different results. Some are better at integrating certain functions than others; some are less expensive in prototyping; others are less expensive in production. Although it seems that many suppliers of processes and design tools try to be all things to all people (claims and counterclaims abound), many such claims breakdown in practical application. Still, there are vendors and customers alike who attempt to force square pegs into round holes. The goal at this stage should be to identify the combination with the best fit for the application.

The ASIC Design Flow

Most ASIC design methodologies will proceed according to some variation of the flow chart illustrated in Figure 2.1. Responsibilities for the various activities will generally be shared between the ASIC vendor and the user. The extent of the responsibilities, interaction, and data exchange will depend on the particular methodology chosen, as well as the selected vendor/user interface. The details of each step in the design flow will be discussed in subsequent chapters, but the flow chart is presented here briefly to provide a top-level view of the generic design flow. The various design methodologies will implement this flow in differing ways, and each will produce differing results.

ASIC Design Methodology Choices and Trade-offs

Design methodology is a rather broad term whose definition encompasses the physical form of the design, as well as the methods employed to describe, capture, and analyze it. At the most basic level, these methods include full-custom layout (handcrafted, all mask layers user-defined), cell-based and compiled-custom (custom and semicustom, standard cell, all mask layers user defined), array-based (semicustom, interconnect metalization mask layers user defined), and user-programmable logic (including field-programmable gate arrays). Figure 2.2 illustrates each methodology and where they fit in the ASIC family tree.

8

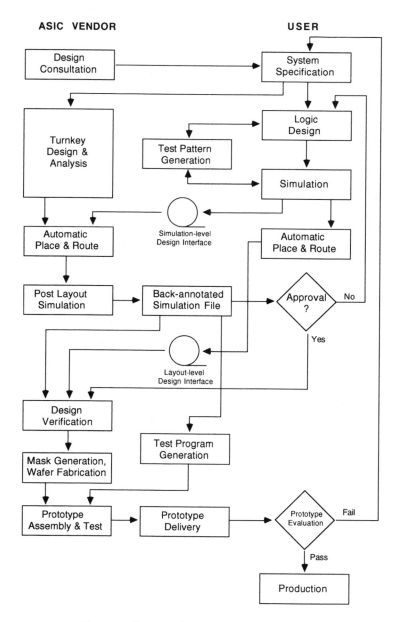

Figure 2.1 Typical ASIC design flows and interfaces.

9

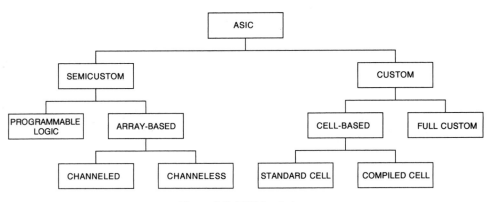

Figure 2.2 ASIC family tree.

Custom

Custom design is so called because each wafer fabrication layer is unique to the particular ASIC design. In contrast, semicustom approaches use predefined cell structures that are layed out on prefabricated transistor grids. Only their interconnections (via metal interconnection mask layers) are required to define the user's design.

Full-Custom

In full-custom design, each circuit element comprising the design (transistors, resistors, capacitors, and element interconnects) are individually drawn and positioned in the circuit layout. For this reason, they are frequently referred to as *handcrafted* designs (although their layout is facilitated by computer-aided engineering tools). One key advantage of this approach is design flexibility. Each circuit element can be optimized for its particular function, and its silicon area can be minimized. The full-custom approach, however, requires highly skilled circuit design and layout personnel, and a single design may take several man-years to complete. Another significant drawback of this approach is that full-custom designs can only be fabricated in the processes for which they were originally designed. That is, full-custom circuit layouts are specific to the design rules (minimum circuit element width and spacing requirements) of the target process and are generally not *portable* to alternate or subsequently more advanced processes. In the event an alternate source is required, the same lengthy and costly design process must be iterated for the new process. The exception to this rule applies to circuits that were designed using layout rules that are so loose that they can be fabricated on a number of different processes—without violating any of their layout rules. The obvious drawback of this approach, however, is that a primary advantage of full-custom design is lost; the circuit is not optimized to take full advantage of a given process' capabilities for performance and area density.

The full-custom design approach is traditionally employed by merchant market IC makers, where high production volumes demand the smallest possible die. As a result, full-custom designs represent the largest production unit volume of all IC design methodologies. But due to time-to-market constraints and high development costs, an increasing number of new and updated standard components are being designed with cell- and array-based methodologies. Consequently, full custom represents the only ASIC design methodology that is declining in use (fewer than 10% of all new ASIC design starts).

Cell-Based

Cell-based design methodologies offer a compromise between full custom and array-based design methodologies. Cell-based approaches offer flexibility in layout but utilize predefined circuit elements called *cells*. A cell may be as simple as a resistor or a gate or as complex as a multiplier-accumulator. The configuration of the cells may be fixed or parameterized, and their placement in the circuit layout is not limited to a fixed grid, as is the case for array-based design methodologies. The design process is vastly simplified when compared to full custom because no special knowledge of the internal transistor-level design of the cells is required. The cells are predesigned and stored in a library that is specific to the ASIC vendor's process. The user simply instantiates the symbols of the desired cells in a circuit schematic, simulates the design, and then forwards the design database to the vendor, who performs automated, computer-based physical layout.

Cell-based and compiled-custom methodologies are yielding designs that are beginning to rival those of full custom in terms of area density and performance, but are developed in a fraction of the time and at significantly lower cost. Although the ASIC market continues to be dominated by array-based methodologies, cell-based and compiled-custom approaches are rapidly gaining in market share.

Cell libraries and module generators are exhibiting more complex, higher-level building blocks (predesigned and characterized cells or macros). They permit the integration of core microprocessors and peripheral controllers, RAM, ROM, mixed digital/analog functions, and complex datapath elements (multipliers, register files, adders, and other bus-oriented functions) far more efficiently than their array-based competitors (see Figure 2.3). Historically, though, the availability of cell libraries has lagged array-based offerings due to the all-layer nature of their design. As a result, they have been difficult to port to new processes. A tremendous effort is required to redesign and characterize process-specific cells in a manner that takes advantage of the new process's capabilities. Process-independent design methodologies (compiled custom), however, have relieved much of this burden. Whole libraries can now be migrated semiautomatically, requiring only design rule checks and corrections and qualification of the simulation models.

12

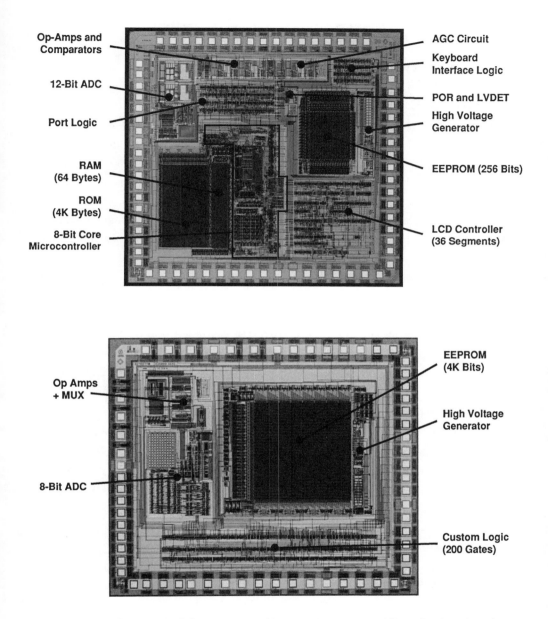

Op-Amps and Comparators

12-Bit ADC

Port Logic

RAM (64 Bytes)

ROM (4K Bytes)

8-Bit Core Microcontroller

AGC Circuit

Keyboard Interface Logic

POR and LVDET

High Voltage Generator

EEPROM (256 Bits)

LCD Controller (36 Segments)

Op Amps + MUX

8-Bit ADC

EEPROM (4K Bits)

High Voltage Generator

Custom Logic (200 Gates)

Figure 2.3 Cell-based ASIC chip designs. (Photographs courtesy of Sierra Semiconductor)

13

The market growth of cell-based/compiled-custom, though, has been impeded by its somewhat higher NREs (nonrecurring engineering or cost of producing prototypes) and longer lead times than for array-based methodologies. This is due to the differences in the manufacturing requirements between cell- and array-based ASICs. The fabrication of cell-based/compiled-custom designs involves customization of all fabrication steps (typically 12 or more mask layers for a standard CMOS process). Conversely, array-based ASICs are fabricated on preprocessed wafers that require only the interconnect layers to be customized (two or more mask layers, depending on the number of metal interconnect layers). However, as designs become more complex, the NRE cost gap narrows. The NRE costs for a 10,000-gate design, for example, may be twice as much for the cell-based approach as the equivalent gate array implementation, but the prototype costs for a 50,000- to 100,000-gate design will be nearly equal for both cell- and array-based approaches. This is due to the magnitude of the layout, simulation, and design verification tasks for such large designs (the associated costs begin to swamp the costs of the cell-based approach's extra mask layers).

Regardless of the design complexity, the cell-based implementation will generally yield a smaller die size than would an array-based methodology. Put another way, the cell-based approach may allow far more design integration than an array of the same size. This is so for several reasons. First, cell-based layouts are not restricted by the array's universal layout grid. Since they are customized on all layers, they don't have to conform to the pattern imposed by generic, preprocessed array structures. Also, where transistor sizes are fixed in arrays, they may be optimally sized in cell-based layouts. This leads to tighter packing, shorter interconnects, and higher performance. For the same reasons, memory is significantly more area efficient in cell-based designs as optimized structures can be defined. Also, the aspect ratio (x, y dimensions) of the cell-based chip can be varied, providing more flexibility in packaging. Arrays, on the other hand, are typically square and their dimensions are fixed.

The substantially smaller die sizes enjoyed by cell-based ASICs produce another benefit in the area of production unit costs. Although the NRE charges may be somewhat higher than arrays, the production unit costs may be as little as half that of the equivalent array-based design. Depending on the production volume, cell-based ASICs can provide a far more cost-effective solution.

In the case of pad-limited designs, however, the advantages of cell-based/compiled-custom methodologies are less obvious. Pad-limited designs result when the number of input, output, power, and ground pads in the periphery of the chip determines the minimum die size. In such cases, cell-based designs offer no die size advantage. This assumes, of course, that the array and the cell-based chip have the same pad pitch (pad-to-pad spacing). Still, a production cost advantage may be enjoyed since the core area of the cell-based chip (area of the chip occupied by active circuitry) will be smaller than that of the array-based design. This should result in better manufacturing yields and, therefore, lower costs. Figure 2.4 illustrates both pad- and core-limited layouts.

Since cell-based ASICs are less expensive in production quantities, gate array-based designs are frequently converted to cell-based when production volumes reach the economic breakeven point. Since many gate array suppliers also support cell-based design, the conversion is fairly straightforward. The netlist (description of all cells in the design and their connectivity) is identical, and the cells are compatible or roughly equivalent. Figure 2.5 illustrates a sample circuit and its netlist. Furthermore, the standard cell's increased density, relative to gate arrays, can improve   overall performance.

So far, we have described the distinction between array-based and cell-based/compiled-custom in terms of their manufacturing differences. In this context, cell-based and compiled-custom designs are very similar; both require customization on all mask layers and fabrication process steps. There are, however, significant differences between cell- or standard cell-based designs and compiled-custom designs.

In standard cell design, the circuit is implemented using a fixed library of primitive functional elements. These include basic logic gates (AND, NAND, OR, NOR, XOR), complex gates (AND-OR-INVERT), flip-flops, latches, and so on. The library may also include higher-level functions, such as fixed-bit-length counters, shift registers, ALUs and adders that are composed from the primitive elements.

Compiled-custom designs, however, are not limited to a set of predefined cells. Rather, they employ function generators that permit the generation of cells and

15

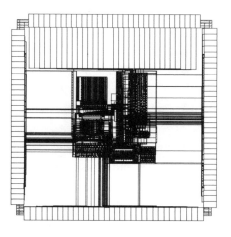

a

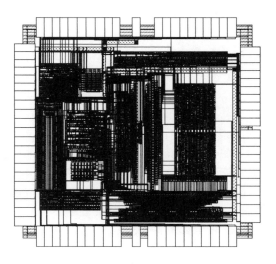

b

Figure 2.4 Pad-limited design (a) versus core-limited design (b). (Provided courtesy of Seattle Silicon and Cascade Design Automation)

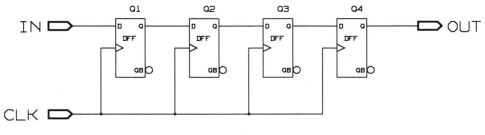

a

MODULE COUNTER;

INPUTS   CLK IN;
OUTPUTS OUT;

LEVEL FUNCTION;
DEFINE

CLK = (CLK);
IN_NET = (IN);
OUT_NET = (OUT);

Q4 (OUT_NET=Q) = DFF (CLK=CLK, Q3Q1=D);
Q3 (Q3Q1=Q) = DFF (CLK=CLK, Q2Q1=D);
Q2 (Q2Q1=Q) = DFF (CLK=CLK, Q1Q1=D);
Q1 (Q1Q1=Q) = DFF (CLK=CLK, IN_NET=D);

END MODULE;

END;

b

Figure 2.5 Counter schematic (a) and corresponding netlist (b).

17

supercells which are compiled according to the specific requirements of the design. The user effectively creates his own custom library of cells for use in a given project.

Cell, module, or function (the terms are used interchangeably) generators (or compilers) are the result of very sophisticated software algorithms that automatically configure the physical layout of a cell in accordance with a specific set of process design rules. Design rules essentially define minimum width and spacing requirements for fabricating transistors and interconnections in a given process. Depending on the supplier, the cell compilers may be process specific or process independent.

When a cell is created, the compiler software generates all the necessary design views. These include the physical layout geometry (expressed in the target process's design rules), a simulation model (gate and/or transistor level), and a symbol, for use in schematic capture.

The compilation process begins with a basic cell description, which is parameterized for customization by the user. Parameters include such attributes as bit width (for counters and registers, for example), presence or absence of preset and clear lines, RAM organization, and so forth. This basic library will typically include everything generally available in a standard cell library plus a number of higher-level cells, or macrofunctions. This capability frees the user from limited, fixed, low-level libraries and allows design capture at higher levels of abstraction—a benefit that greatly enhances productivity. Leveraged by these higher-level design techniques, weeks and even months of design time can be trimmed from the development schedule. In contrast, gate array and standard cell-based design (based on a low-level fixed library) requires the user to manually partition his design to the gate level.

The tools used to generate compiled-custom layouts are typically referred to as silicon compilers. ASIC suppliers (who support cell-based design) are using these automatic cell generators as a means of extending their libraries and for offering special, custom functions (specific to their process) to their customers. This is a strategic capability since standard cell libraries can never contain every possible cell the user may need. There are, however, some drawbacks associated with this capability. A cell compiler has so many possible parameter combinations that the vendor cannot possibly generate and characterize every combination. Therefore, characterization data may be based on interpolation rather than direct measurement.

18

Where compilers are not available, higher-level functions, or macros, are composed by interconnecting lower-level elements in the cell library. For example, building an 8-bit counter entails connecting eight D-flip-flops along with any desired control logic. The disadvantage of this approach becomes evident when the macro cell is laid out with the rest of the chip. The macro cell gets decomposed into its constituent elements, which may then be dispersed over the entire chip to obtain an efficient layout. Since the cells making up the macro function are not grouped together, its timing may be difficult to predict. Due to the soft nature of their layout, they are commonly referred to as soft macros. This is true of any design that is implemented with low-level cells. The place and route software will arrange and rearrange the cells as required in order to obtain a good layout or to fit into a particular array.

A few companies offer silicon compilation tools in vertically integrated ASIC design systems that include parameterized module libraries as well as tools for schematic capture, simulation, timing analysis, and layout. They will typically include a number of process design rules that allow the user to evaluate his design in any number of them for performance, size, and cost trade-offs. In this way, the optimum process can be determined before any commitments to a silicon supplier are made. Compilers are also the enabling technology for migrating designs as well as entire libraries to newer, more advanced or alternate processes without an extensive redesign effort.

Although compiler tools are generally process independent, they will still require calibration to the desired process. Even though some processes may be compatible with others, a set of generic design rules that can be fabricated universally does not exist.

Process qualification can be an expensive and lengthy undertaking. The net result, though, are compilers that are *capable* of generating layouts without errors (design rule violations) and simulation models that accurately reflect process performance. Suppliers of silicon compilation tools, as a practical matter, prefer to limit the number of available qualified processes. A smaller number is simply easier to maintain. Consequently, the selection of processes is limited and the actual degree of process independence is somewhat constrained.

19

Array-Based

Array-based methodologies represent, by far, the largest ASIC market. Roughly half of all mask-defined ASICs are array-based. Their popularity is due primarily to their lower NREs and shorter fab cycles (relative to cell-based options).

Gate arrays are preprocessed up to the final metal interconnect layers, which customize or personalize the array. These metalization patterns wire up the macro cells and complete the routing connections between them. The array bases, or master slices are fabricated in large quantities, so their one-time mask costs are shared by all designs processed on them. The benefits to the user are lower NREs (two or more additional mask steps are required to complete processing) and faster turnaround for both prototypes and production. Another benefit afforded by the gate array master-slice methodology is faster production ramp-up and generally shorter lead times, since master slices are prefabricated and maintained in inventory.

There is often a need to quickly revise an ASIC design due to a system specification change or design error. Such changes require prototype (mask generation, fabrication, assembly and test) iteration. As a result, there is a schedule and cost penalty for each iteration incurred. Relative to all-layer, cell-based approaches, the penalties for array-based design iterations have far less impact.

Gate Array Architectures

Array-based ASICs fall into two basic categories: channeled, or conventional gate arrays, and channelless, or sea-of-gates arrays (Figure 2.6). Channeled arrays are so called because an empty channel of silicon separates rows of unwired transistor pairs. The transistors are wired into gates, flip-flops, and larger functions (per the design netlist) and then interconnected, or routed using the dedicated routing channels. These arrays dominate the market for designs up to 20,000 gates. Beyond that, more efficient layouts are achieved with channelless, or sea-of-gates architectures. The 20,000-gate practical limit is imposed because most processes are restricted to two layers of metal interconnect. To accommodate larger designs, either the channel widths must be

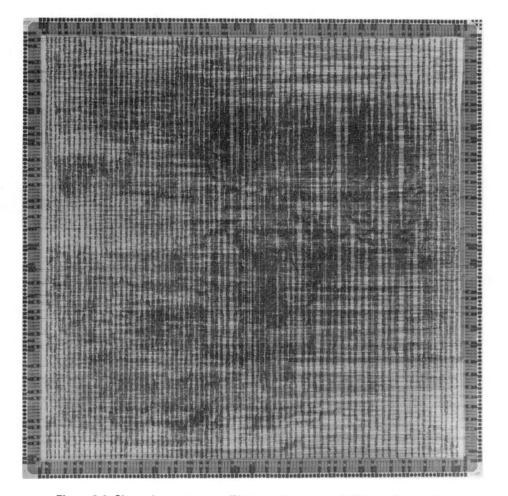

Figure 2.6 Channeless gate array. (Photograph courtesy of LSI Logic Corporation)

21

increased or additional interconnect layers must be added. Increasing channel widths is unacceptable because it results in a larger die. Adding additional levels of interconnect, though, reduces overall die sizes or, for the same area, allows an increase in the possible level of integration.

While three- and four-layer metal processes are enabling greater densities for channeled arrays, channeless or sea-of-gates arrays are raising the ceiling on raw gate count. In sea-of-gates arrays, the channels are removed. The entire array is covered with potentially active and usable transistor cells, or a sea of gates. In the sea-of-gates architecture, unused transistor sites serve as routing channels. The interconnect metal is deposited directly over the unused transistors.

Although they have been criticized for their relative silicon area inefficiency, they are able to raise the absolute gate count far beyond the capability of channeled arrays. In this sense, 40% utilization of the 100,000 or so *available* gates is better than 95% utilization of 20,000 gates. But like conventional channeled arrays, sea-of-gates arrays also benefit from the addition of three- and four-layer process technologies, which will significantly improve routability and utilization. Three- and four-layer processes, however, require additional mask steps, which increase costs and may contribute to manufacturing yield loss.

Counting the Gates

Gate arrays are available in a range of discrete sizes. The complexity, or equivalent gate capacity, generally increases in approximately 35% steps. Because of this granularity, it is unlikely that a particular design will exactly fill an array. Matching the design to the most appropriate array size, therefore, can be tricky. If the design doesn't fit, the user is forced to move up to the next higher (and more expensive) array size.

Gate array capacities are generally measured in terms of total equivalent or available gates. An equivalent gate is generally defined as a two-input NAND gate. Since gate arrays are essentially transistor arrays, many array manufacturers calculate the total number of transistors on the array (including those that will be consumed by routing) and then divide them by four—the typical number of transistors per gate. Other vendors base the equivalence on the number of available transistor cells or groups of transistors that are electrically isolated from neighboring groups. Not all the transistors in a cell may be usable with some functions. It is also important to note that macro cells from different vendor libraries may use different numbers of equivalent gates for the same function.

An accurate evaluation of a gate array's capacity must consider more than the number of theoretically available raw gates. The degree of actual utilization is extremely design dependent. For a designer to know if a design will fit into a given array, some level of benchmarking will be required. An appropriate benchmark will take into account the mix of functions that are typical of the designer's applications. For example, some applications contain a great deal of random or combinational logic, while others are more structured and regular. Due to differences in gate array architectures, the manner in which these circuits are placed and routed on the array will vary significantly. For example, a design may map nicely into a 10,000-gate array from vendor A, but may require moving up to a 12,000-gate array from vendor B. Likewise, a highly structured circuit design might fit into a 7,000-gate array, but a design consisting of combinational logic, with an equivalent number of gates, might require a 10,000-gate array. Typical designs, though, will contain a mixture of regular and random functions. The utilization, then, will be determined by the ratio between the two circuit types.

Routability is the primary limiting factor in gate array utilization. Because all circuits are not equally routable, several benchmark circuits are needed. Also, certain circuit designs, such as those consisting primarily of random logic or designs containing multiple global buses, are more prone to routing congestion. Routability, therefore, may be the more meaningful metric of array capacity. A typical benchmark strategy is to design a set of appropriate benchmark circuits and then map them repeatedly into the benchmarked arrays until they are consumed.

Estimating Performance

The performance of a gate array design is difficult to predict prior to routing. The cell placement and routing results can vary greatly from design iteration to iteration. Generally, the placement and routing operation is performed by the ASIC vendor.

23

Although the user may be able to specify a few critical paths, the vendor will generally have no detailed understanding of the design specification. Thus, the user loses considerable control in this very important design aspect.

Once the design is placed and routed, the database is returned to the user, with the actual post-routing delays back-annotated to the simulation files. The user can then verify the integrity of the design before committing it to silicon. Although pre-route statistical back-annotation programs are available, an increasing number of users are beginning to perform the placement and routing operations on their own workstations. (See Chapter 5 for a more detailed discussion.)

Integrating RAM

One of the more useful functions to integrate into an ASIC is static RAM. Uses include local memory, scratch-pads, register files and data buffers. Some of the advantages of on-chip RAM include greater performance (through the elimination of I/O buffer delays), potentially lower cost, and lower ASIC I/O count (no need for external address and data pins).

As might be expected, the different ASIC methodologies have different approaches to integrating RAM. Cell-based design is by far the most efficient approach. RAM structures can be optimized (RAM transistor device ratios are different than those for logic) and compiled for the precise organization needed. Furthermore, their aspect ratios (x, y dimensions) can be manipulated to achieve an optimal chip layout.

The inclusion of RAM in gate arrays, on the other hand, is much more restrictive. The channeled array architecture, for example, is at a severe disadvantage. First, implementing a single memory bit in an array may consume as many as six gates. Since the transistor layouts comprising the gates are designed for general-purpose use, they are not optimized for special functions, such as RAM. In fact, gate arrays' fixed transistor sizes are set to a large size to account for worst-case drives. As a result, their power consumption is also relatively high. Second, the routing channels between the cell rows result in considerable wasted silicon area. Finally, this arrangement imposes a significant performance penalty. Although these restrictions may be acceptable for

24

very small RAM densities, for most applications, the cost of RAM integration in such arrays is simply too great.

Structured or embedded arrays (see Figure 2.7) offer a partial solution to this problem. In this approach, an optimized block of RAM (of predetermined size) is prefabricated in an allocated area on the base array. These on-chip RAM arrays are able to support a variety of configurations—up to some specified limit. This approach can only be efficient, however, when the total available RAM capacity of the array is needed. Like unused gates on an array, unused memory is a waste of silicon area.

When evaluating on-chip RAM options, consider the maximum allowable size, the organization flexibility, and the method of configuration and interconnection. If a particular methodology requires you to build a RAM by interconnecting a number of smaller blocks of RAM through the use of decode logic, it will be far less area efficient than a single block. Its performance will also suffer.

When comparing the performance of on-chip RAM with standard, off-chip RAM products, remember to take the chip-to-chip delays into account. These will be eliminated in the on-chip partition.

Programmable Logic

For high logic densities or high speeds, a masked-programmed ASIC may be the only solution that meets the design requirements. But recent and continuing advances in programmable logic devices (PLDs) have made them suitable alternatives to some mask-programmed ASICs. If the application requires fewer than 5,000 gates and the I/O needs are relatively low (fewer than 100 pins), then programmable logic, particularly in the form of field-programmable gate arrays (FPGAs), may be appropriate.

Programmable logic represents a low-risk entry point into the benefits of ASIC technology. As such, they are also the fastest growing segment of the ASIC business. Their low risk and increasing complexity have helped FPGAs encroach upon the low end of the gate array market—particularly in low-volume applications. When one considers that the median ASIC equivalent gate count is approximately 15,000, its easy to see that gate arrays are rapidly becoming vulnerable to FPGA developments.

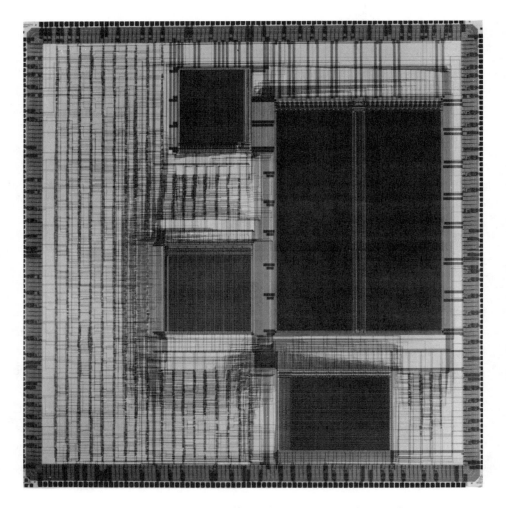

Figure 2.7 Embedded memory array. (Photograph courtesy of LSI Logic Corporation)

Programmable Logic Is Low-Risk

Much of the risk associated with ASICs results from design errors, specification changes, incomplete original specifications, and processing mistakes. In contrast to mask-programmed ASICs, FPGAs are very forgiving and are able to alleviate these risks in a number of ways:

- Design changes can be implemented quickly and easily.

- Performance is fairly easy to predict.

- System-level verification can begin much sooner.

- Rapid time to market.

- Short production order lead times.

- Flexibility in order quantities.

- Multiple sources.

- Distributor stocking.

FPGAs offer the user significant advantages in cost and time to market. They can be ordered, received and programmed in a matter of days, compared with months for mask-defined ASICs. Also, respecting time to market considerations, it may make sense to use FPGAs to make a product introduction deadline, while developing the mask-programmed version in parallel.

FPGA Architectures

Selecting an appropriate programmable logic device can be a complicated matter. Designers must consider not only device density but also performance, design tool support, packaging options and cost. The choice is made even more complex by the number of architectures available.

All FPLDs fit into one of two categories: programmable logic devices (PLDs) and field-programmable gate arrays (FPGAs). Each category features a variety of device architectures. PLDs consist of AND-OR array planes that feed into flip-flops and have a fixed interconnect architecture. Devices in this category include PALs, FPLAs, EPLDs, and EEPLDs. FPGAs, on the other hand, feature a flexible interconnect technology with no fixed AND-OR planes. Instead, they incorporate an array of identical logical building blocks embedded in a matrix of programmable and fixed interconnection lines. Each building block, or module, contains flip-flops and combinational logic that can be configured (via user programming) in a number of ways and can used to create higher-level internal functions like counters, latches, and shift registers. Because the interconnects run along paths between the modules, their structure resembles that of a conventional channeled gate array.

The FPGA I/O cells are also programmable and typically contain flip-flops to latch input and output signals. Some devices allow output registers to be used as buried registers, freeing up associated I/O for other uses. Each device output is a function of a limited set of device inputs. Because of these limitations, multiple devices often have to be cascaded to implement functions that require wide gating, causing some degradation in performance. As a result, FPGAs are better suited for register-intensive, narrow gate functions.

With FPGAs, each manufacturer uses its own architecture and functional logic module to implement various circuit functions—and they don't all count gates in the same manner. Furthermore, the way the building blocks are used in a design determines how many of the gates can actually be used to implement logic.

Architectures also vary in terms of their circuit technology. Devices may be based on static RAM, EEPROM, and fused or anti-fuse links. For those devices that feature a static-memory-based architecture, logic changes can be made by reconfiguring the programmable logic right in the system. Configuration of SRAM-based devices is accomplished by programming the internal static memory cells that determine the logic functions and interconnections. The configuration programs can be down-loaded to the devices at system power-up or by command.

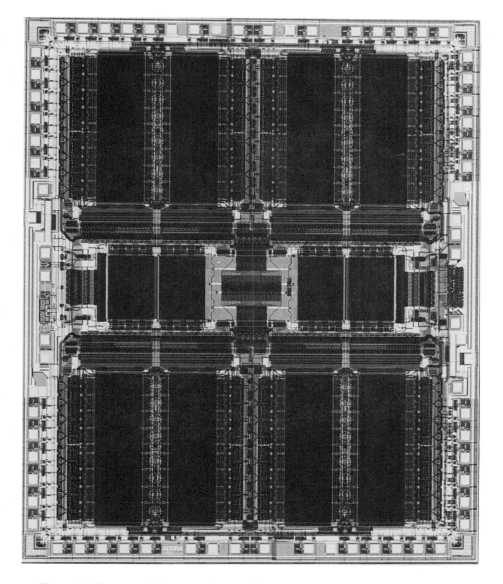

Figure 2.8 Programmable macrocell array. (Photograph courtesy of Altera Corporation)

Evaluating PLD Performance

The process of evaluating a new technology typically begins with the device data sheet. But even here, the user must know how to interpret the specifications. Also, the device must be evaluated in the context of the application.

Although conventional PLDs are very fast, their logic density is low (generally less than 1,000 gates). FPGAs, however, are much larger, but are generally much slower than PLDs.

Toggle rates as specified are essentially meaningless. In actuality, FPGAs run at a fraction of their stated toggle rate. For some FPGAs, the term 100 MHz device is used to designate a device with a flip-flop toggle frequency of 100 MHz. Unfortunately, this specification is not an accurate measure of true in-system performance. Logic interconnect delays and the number of logic levels needed in a given path can reduce the effective performance of the device to one tenth of its stated performance rating. The factors that determine real performance include input delays, setup and hold times, logic module propagation delays, interconnect delays (module to module), output delays, and the number of logic levels required. The logic module delay is multiplied by the number of logic levels needed. Also, devices are typically cascaded, further increasing delay. The sum of delays for critical paths determines the actual operating frequency for the device.

The quality of the design tools can also affect device performance. To yield optimum integration and performance, the tools must incorporate state-of-the-art logic synthesis and place and route algorithms. The tools should also allow the user to specify and optimize critical paths.

Again, like other ASIC methodologies, the best way to evaluate the manufacturers' data sheets is to benchmark their devices. Benchmarks should include both registered and combinational logic functions. The user should strive for full device utilization in the benchmark as placement affects device performance.

Cost Effectiveness

The economic feasibility of the programmable logic approach should be considered carefully if any significant production volume is anticipated. The comparison of costs among the various methodology alternatives should include not only the unit cost of a device, but the amortization of all engineering costs over the total anticipated production volume as well.

The cost effectiveness of a logic device is largely determined by its cost per gate. FPGAs come at a much higher cost per gate in production when compared with conventional mask-programmed ASICs.

A device's gate capacity boils down to the amount of logic that can be realized from the logic modules and the number of modules that can be used. Gate counting alone, however, does not provide an accurate picture of device density. For example, a device claiming a 5,000-gate density may actually have only 1,500 usable gates. Like gate arrays, FPGA vendors don't all count gates the same way. For FPGAs, the device architecture is the more important consideration, and benchmarking is the only way to separate the wheat from the chaff.

FPGAs will generally be the most cost-effective ASIC solution for gate counts of 5,000 or less and production quantities in the hundreds. Gate arrays generally have a cost advantage once production reaches the 1,000-piece level, regardless of the gate count. The production economies of scale afforded by array- or cell-based ASICs, though, eventually attract many FPGA-based designs. For example, a single $10 gate array might replace as many as five $5 FPGAs while also reclaiming valuable board space. The comparison, however, must also include the NRE and schedule costs. Also, the conversion from multiple FPGAs to a single ASIC may pose significant problems and hidden costs. These would include mapping the design to a gate array and consolidating the associated test programs.

Finally, due to rapidly changing markets and other business considerations, fewer than half of all ASIC design starts mature to production. When FPGA-based designs do not go into production, the cost consequences are far less than those for mask-defined ASICs.

31

FPGA Development Tools

Development tools run the gamut from sophisticated systems running on workstations to simple low-, or no-cost logic compilers that run on a PC. Some FPGAs, though, may require special software that may cost thousands of dollars.

The increased cost of FPGA development tools over those for standard PLDs is due to the more complex nature of the FPGA devices. The design tools and methods used for FPGAs are significantly different than those for conventional PLDs. Unlike PLDs, FPGAs require that the logic be placed (assigned to logic modules) and routed between the individual logic modules of the array. Depending on the architecture, the tools may be specific to the device and offered only by the device supplier.

More generic development tools work with a large variety of devices. These systems typically incorporate a device library that contains information about the parts' performance, cost, architecture, pin-out, and more. Some will prioritize a list of the most efficient devices for a particular design and may recommend selections based on user-supplied criteria, such as a favored manufacturer. The quality of the selection obviously depends on the completeness of the library and how easy it is to update with new devices.

The methods of design description or design entry for FPGAs can also be very flexible. Designs may be captured via schematic capture, Boolean equations, truth tables, state diagrams, or logic synthesis languages. Once a design simulates properly, the user can check it out in hardware almost immediately. If the design won't fit into the targeted device or doesn't perform as expected, its a simple matter of selecting an alternative device, recompiling the design, and programming it.

Other tools that greatly simplify FPGA design include design partitioners. Partitioning tools are able to automatically partition, into sets of devices, complex designs that will not otherwise fit entirely into a single device.

Design Validation Using FPGAs

Many ASIC and design tool vendors offer their customers a means to design a circuit using an ASIC cell library and then output the design to a data format that can be used to program an FPGA (or multiple FPGAs if the tools support design partitioning across several devices). This can be a very effective strategy for proving the functionality of a design in hardware before committing it to the mask-defined version. Of course, there are shortcomings to this approach in that the timing may be substantially different than the ASIC version (and its simulations). Still, the low-speed validation of the logic can be very useful. Conversely, the same tools can be used to migrate   FPGA-based designs to vendor-specific, mask-defined ASICs. (For additional information, refer to Chapter 5, subsection *ASIC Emulation.*)

Parameter	Full-Custom	Cell-Based	Array-Based	FPGA
Multisourcing Difficulty	Highest	Low - Med	Low - Med	Lowest
Development Cost (NRE)	Highest	High	Med	Lowest
Mask Costs	High	High	Low - Med	None
Design Time	Highest	Med	Med - High	Lowest
Redesign Flexibility	Lowest	Low	Low - Med	Highest
Cost of Iteration	Highest	High	Med	Lowest
Production Unit Cost	Lowest	Low	Med	Highest
PCB Costs	Lowest	Low	Low	High
Layout Efficiency & Flexibility	Highest	Med - High	Low - Med	Lowest
I/O Flexibility	Highest	Med - High	Low - Med	Lowest
Level of Integration	Highest	High	Med	Lowest

Figure 2.9 Comparative summary of ASIC design methodology options.

Chapter 3

Selecting the ASIC Technology

ASIC TECHNOLOGY REFERS to the fabrication process that is used to realize a design. For each process type, the various design methodologies may be applied. For example, full-custom, cell-based, array-based and programmable logic-based ASICs are all available in CMOS, bipolar, and GaAs process technologies. As is true for the various ASIC design methodologies, each process technology has its place. The selection of the appropriate technology is extremely dependent on the design goals for the particular application. The material in this chapter provides an overview of the primary process technology classes and their dominant attributes, advantages, problems and special considerations.

CMOS

CMOS process technology, thanks to its low power, ease of design, and ever-improving performance, is the most popular of all ASIC technologies. At least two-thirds of all new ASIC designs starts employ CMOS, and it is expected to remain the technology workhorse through the 1990s. Nearly all the characteristics that make CMOS so attractive today are getting even better. As fabrication technology improves, CMOS processes are yielding smaller feature sizes and increased levels of performance. CMOS circuitry is more readily scalable than bipolar processes, due to its simpler gate construction. That is, the process design rules can be scaled down or shrunk, thus enabling higher performance and area density. Also, increasing circuit density does not complicate the power dissipation problem as severely as it does for bipolar. As a result, CMOS lends itself to *very* large scale integration.

With feature sizes shrinking to 0.5 μm, CMOS performance is rapidly encroaching on applications traditionally reserved for high-speed bipolar technologies. As it develops, it is serving high-end, low-end, and all points in between across all design methodologies in a growing number of applications.

In addition to the tremendous improvement in performance, the migration to finer geometries results in a significant reduction in die area. Coupled with increases in wafer sizes, the economic impact is equally impressive.

The addition of third and fourth layers of interconnect metal also affects both density and performance. Performance will improve due to the overall shorter interconnect wire lengths and the lower associated parasitic capacitance. Density will increase, particularly in the use of channeless gate arrays, where usable gates will increase from 50% to more than 90% of the total available.

CMOS is also very versatile in terms of its operating voltages, which range from less than 3 V (battery operation) to 10 V for silicon-gate CMOS. Older metal gate processes can operate up to 15 V.

CMOS processes are available in many forms from a great many suppliers. The primary form is known as *bulk* CMOS. Bulk CMOS is so called because the transistors are formed in the substrate material itself. In variants of CMOS processes, such as silicon on sapphire (SOS) or silicon on insulator (SOI), the transistors are formed in the grown layer of silicon on top of the substrate. These processes exhibit improved latch-up immunity and better radiation hardness over bulk CMOS processes, but are more expensive, and their availability is somewhat limited.

Process Design Rule Considerations

Processes are generally characterized by the size of the transistor gate length. For example, a 0.8μm process rules have a drawn gate length of 0.8μm. The actual processed feature size may be smaller. Generally, the smaller the gate length is, the better the density and performance. But just as important as transistor sizes are the process design rules that specify the metal pitches at each layer. These define the minimum wire widths and the minimum allowable spaces (isolation distance) between two adjacent wires.

For example, Figure 3.1 illustrates the impact on chip size that differences in metal pitches have on a 10,000-gate design fabricated in two different 1.0μm CMOS processes.

Parameter	Process A	Process B
Gate Length	1.0μm	1.0μm
Metal 1 Pitch	2.50μm	3.50μm
Die size (sq. mils)	40,000	68,000

Figure 3.1 Metal pitch comparison.

Although both Processes A and B are designated as 1.0μm processes, the effective densities are dramatically different due to the other limiting design rules. Figure 3.2 illustrates the many variables that characterize a typical CMOS process.

Step Coverage

Although the resolution of available process photolithography technologies has historically limited feature sizes, limits imposed by physics place lower limits on feature sizes as well. The phenomenon of metal migration, for example, places a lower limit on metal pitch (wire width, specifically). If the wires are too small, the high current density (relative to the wires' current-carrying capacity) causes an accelerated movement of the metal atoms until an open circuit eventually results.

Rather than compacting wires (shrinking the metal pitch) any further, newer processes are increasing circuit density by stacking additional layers of metal interconnect wire (Figure 3.3). There are, however, other problems that limit the metal pitches in these additional layers. With each successive process layer, surface irregularities increase and the deposition surface becomes uneven. The vertical steps the metal must cover when it goes into and out of contact holes and over other metal lines become the weakest part of the metal (the metal lines thin out at the knee of the step).

Although the step coverage problem is manageable with two levels of metal, other processing techniques are required for fabrication of additional layers. Newer processes use planarization to smooth the surfaces between layers. Planarization has the effect of softening abrupt steps in the underlying layers, providing a smoother surface for subsequent metal depositions.

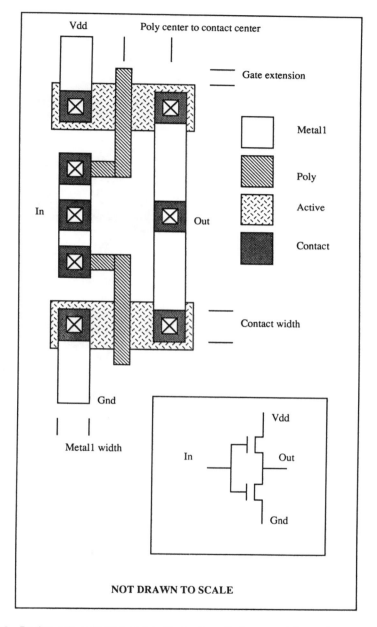

Figure 3.2 Design rule variables and typical values (facing page) for a 1μm process. The example features the layout of a simple inverter circuit. (Illustration provided courtesy of Seattle Silicon and Cascade Design Automation)

RULE	DIMENSION (μm)	DESCRIPTION
Active Area		
aw	3.00	Active area width
ngw	2.00	Minimum n-channel gate width
pgw	2.00	Minimum p-channel gate width
as	3.00	Active area spacing (n+ to n+ or p+ to p+)
Poly		
pwy	1.00	Poly width
ngl	1.00	Minimum n-channel gate length
pgl	1.00	Minimum p-channel gate length
pys	2.00	Interconnect poly spacing
gts	2.00	Gate poly spacing
pyxna	1.00	Gate extension beyond n active area
pyxpa	1.00	Gate extension beyond p active area
axpy	2.00	Drain/source extension beyond gate
pyas	0.00	Poly to unrelated active area spacing
Contact		
cw	1.00	Contact width
cs	1.00	Contact spacing
acgts	1.00	Active area contact to gate spacing
pycas	1.00	Poly contact to active area spacing
aoc	1.00	Active area overlap of contact
pyoc	1.00	Poly overlap of contact
Metal-1		
mw	2.00	Metal-1 width
ms	2.00	Metal-1 spacing
moc	1.00	Metal-1 overlap of contact
Via		
vw	1.00	Via width
vs	1.00	Via spacing
m1ov	1.00	Metal-1 overlap of via
vas	1.00	Via to active area spacing
vpys	1.00	Via to poly spacing
vcs	1.00	Via to contact spacing
Metal-2		
m2w	2.00	Metal-2 width
m2s	2.00	Metal-2 spacing
m2ov	1.00	Metal-2 overlap of via

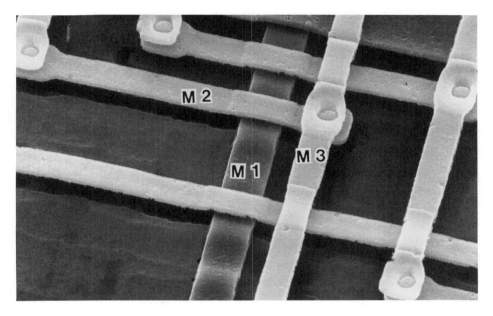

Figure 3.3 Triple layer metal process. (Photograph courtesy of Motorola Inc., ASIC Division)

Performance

With respect to performance, the important design rule considerations are oxide thickness (the thinner, the better), the effective channel length (the shorter, the better), and the overall density (the smaller, the better).

BiCMOS

Although the market for CMOS ASICs is strong and maintaining its market share status, applications calling for BiCMOS technology represent the fastest growing segment. Utilizing the best of both bipolar and CMOS worlds, they excel in applications where high density, low power, high performance, and high drive all carry equal weight. Typical applications include high-end PCs and workstations, telecommunications, and mixed digital/analog.

The combination of the two technologies produces a speed/power performance quotient that cannot be achieved by either process alone. ASICs employing this mixed technology generally consist of a submicron CMOS core with a bipolar I/O pad ring. The CMOS circuits comprising the core contribute high functional density and relatively low power. The bipolar transistors, making up the periphery of the chip, provide high-drive capability as well as compatibility with ECL logic levels.

BiCMOS's high output drive capability (greater than 50 mA) makes it particularly well suited for high-speed, high-current drive applications, such as driving large buses or backplanes. In contrast, the performance of CMOS outputs degrade significantly with increased loads. The high-speed I/O provided by the bipolar pad ring boosts operating frequency 50% or more over a CMOS-only implementation with similar feature sizes.

Load-driving Capability

Although CMOS transistors operate faster as they get smaller, their ability to drive interconnect capacitance suffers. In submicron processes, the intrinsic gate delay is secondary to the delay caused by fan-out loading and interconnect capacitance. For this reason, some BiCMOS processes employ a hybrid CMOS/bipolar circuit technology in the internal logic elements as well. In such circuit technologies, a typical CMOS NAND gate will be buffered by totem-pole bipolar output, which provides extra load-driving capability. Although the addition of the bipolar output stage increases the intrinsic gate delay, the enhanced load driving capability can improve overall path delays significantly and minimize clock and signal skews.

As mentioned, most of the delay (and power consumption) in CMOS circuits results from the charging and discharging of gate and net capacitances. These delays increase with temperature and worst-case supply voltage. Adding bipolar devices to the gate outputs helps to overcome these problems because they are inherently less sensitive to temperature and supply voltage variations. They are also capable of improving performance without incurring a significant power penalty. The bipolar devices, when in their standby condition, draw no dc current (except for junction leakage current). The bipolar I/O circuitry, however, will draw substantially more power than CMOS.

Ground Bounce

ASICs designed in submicron standard CMOS processes are capable of supporting system speeds in excess of 50 MHz. Enjoying the full benefit of this capability, though, is somewhat impeded by effects such as ground bounce, or transient spikes that occur when a large number of high-drive outputs on an IC switch simultaneously.

40

The sharp rise and fall edges of the CMOS output structures are to blame. Unfortunately, the solutions to the ground bounce problem incur significant cost and performance penalties. Building controlled rise-time (or slew rate) features into the output pads inevitably slows them down. Reducing the supply voltage, another method of reducing switching noise, reduces the speed of the entire chip, not just the outputs. Another common approach to the problem involves placing additional ground pads between the outputs whose states change simultaneously. The obvious drawback to this solution, however, is the requirement for additional pins and therefore, a larger, more expensive package.

BiCMOS, on the other hand, overcomes these difficulties. Although the propagation delay through a bipolar output is much faster than CMOS, its edge rates are slower, thus minimizing, if not eliminating, ground bounce effects in many applications. In the case of ECL-compatible outputs, their 800 mV swing dramatically cuts transient noise effects.

Cost Impact of BiCMOS Processing

BiCMOS is a complex technology from a process manufacturing perspective. The additional mask layers and extra processing steps result in significantly higher costs (and lead times) for both prototyping and production. Extra mask and process steps also have the effect of reducing yield.

Bipolar

Bipolar development, primarily in the form of ECL (a bipolar circuit technique), has not been idly standing by. Its speed/power/density quotient is improving at a remarkable rate such that it will soon be a viable alternative to BiCMOS, particularly in ultrahigh-performance applications. In the meantime, however, its relatively high power dissipation (the transistors are always partially on) has limited its complexity to about the 10,000-gate level. Some vendors are increasing the number of usable gates by scaling down transistor sizes and by adding low- and high-power options in certain array offerings. Of course, the low-power option is traded for performance (which may be three times slower than the standard, high-power option).

A significant disadvantage of ECL is its requirement for more expensive cooling systems. Even though recent improvements in package technology have permitted forced-air cooling of 40 W chips packaged in ceramic with built-in heat sinks,

41

extensive thermal analysis and support engineering are required. Evaluation and use of ECL requires a total-system perspective, which includes reliability, packaging, and cooling. The cost and size of power supplies must also be taken into account.

The amount of logic on a particular ECL gate array that can be used depends on the number of gates operating at full speed. In some applications, it may be only a fraction of the available gates. In most cases, if the equivalent gate count is multiplied by the power dissipation per gate, the total power will exceed the data sheet's maximum rating. Also, the published maximum clock frequency may be realizable only when a maximum of 10% of the available gates are used.

In spite of its shortcomings, ECL offers greater density than GaAs and greater speed than BiCMOS. Also, in particular contrast with CMOS, the propagation delays of ECL macros change very little with increased loading.

GaAs

Gallium arsenide (GaAs), though it represents a comparatively small percentage of the ASIC market, has the potential of becoming a mainstream ASIC technology for ultrahigh-performance applications. Most GaAs applications are found in low-volume military, fiber-optic, telecommunications, and radar systems. For a variety of reasons, however, its acceptance has been limited. These reasons include its relatively low levels of integration, small die size limitations, processing difficulties, lack of design tools, cost and reluctance to move to a new, little understood technology. Continuing improvements in manufacturing techniques, though, are allowing for significant advances in density and lower costs. Still it is expected to remain a niche process.

GaAs Technology

The GaAs logic gate is a relatively simple structure. A two-input NOR gate requires only three transistors and no resistors. In contrast, a basic ECL NOR gate uses six transistors, three resistors, and several voltage references. This difference in simple gate complexity means that a GaAs circuit will occupy about half the area required for the equivalent ECL circuit.

Most GaAs circuits operate from a single -2 V power supply. This allows performance that is comparable to ECL but at one-half to one-fourth the power. Because they dissipate much less heat, GaAs ASICs require less severe cooling measures typically associated with ECL. Also, GaAs operates well at higher temperatures.

Because GaAs is intended to compete directly with ECL, most GaAs array I/Os can be programmed to be compatible with 100K, 10K, or 10KH ECL voltage levels, thus easing its integration into ECL-based systems.

The GaAs process is fairly simple, relative to other technologies. A typical GaAs process may require as few as 9 masks, where a BiCMOS process may need as many as 24. One would think, then, that GaAs would be a relatively inexpensive process to manufacture. But due to the high cost of the base wafers, small wafer size, and high manufacturing defect levels, GaAs devices can actually be more expensive. A complete cost analysis, though, must also consider the lower package, power supply, and cooling costs compared with competing ECL technologies.

The ultrahigh speeds made possible by GaAs present problems as well as solutions for many applications, particularly at the package and board levels. At higher speeds, the package becomes a limiting factor. With operating frequencies in the gigaHertz range, package lead capacitance and inductance have a significant impact on performance.

Another attraction of GaAs is its inherent radiation hardness. Although GaAs is rad-hard to the total dose level of 20 Mrad, it is still vulnerable to single-event upset (SEU), or soft errors. Special circuit techniques alleviate this susceptibility, but test data (supplied by the foundry) should be carefully evaluated.

GaAs on Silicon

The brittleness and warping of bulk GaAs wafers have limited wafer diameters to 3 to 4 in. New processes are emerging, however, that grow epitaxial GaAs layers on silicon wafers. It is hoped that the less fragile, larger-diameter silicon wafer substrates will yield the economies of scale associated with silicon processing.

Many problems are still to be solved before GaAs on Si can move into mainstream production, though. These include both lattice and thermal mismatches between the two materials. Also, there are processing incompatibilities. GaAs is processed at much lower temperatures than silicon and uses different metalization systems. Research is continuing and alternatives such as sapphire substrates, which provide a better thermal match, are being explored.

Radiation-hardened Processes

Many device properties are affected upon exposure to the radiation that results from a nuclear blast or long-term exposure to lower levels of radiation in space applications (see Figure 3.4). The environments and effects that are produced include electromagnetic pulse (EMP), ionizing dose rate, total dose, and neutron fluence. These different types of radiation have different effects on the device. From a systems level, the effects can include a recoverable upset, data loss, incorrect operation, degraded system performance, and even permanent damage.

EMP Effects

EMPs can result from high-altitude nuclear explosions that produce a high-intensity electromagnetic field. Semiconductors may become functionally upset, fail, or incur permanent damage resulting from the current transients induced by the EMP. The principle EMP-hardening technique involves shielding the system from the transient by attempting to reduce its magnitude before it reaches sensitive electronics.

Ionizing Dose Rate

The ionizing dose rate (measured in rads in silicon per second) is the rate at which the nuclear by-products (gamma rays and x rays) bombard the semiconductor material. The threshold of effects in semiconductors is $>10^5$ rads (Si)/s. The effects include development of a photocurrent, which can lead to latch-up and device burnout. This condition can generally be prevented by powering down the device (upon detection of a radiation environment) or by current-limiting all paths to ground, using resistors.

Total Dose

Total dose (measured in rads in silicon) refers to the total amount of ionizing radiation from all sources. The excess charge that results can become trapped in the gate oxide material. The threshold of effects in semiconductors is $>5 \times 10^2$ rads (Si).

The extent of device degradation depends on the quality of the oxide material (the thicker the oxide, the greater the volume available for charge buildup), as well as the electrical bias during irradiation. Special, low-temperature processing techniques are able to enhance the total-dose radiation hardness of the gate and field oxides.

Total Dose, rads (Si):	100	1K	10K	100K	1M
Nuclear Explosion			▶▶▶▶▶▶	▶▶▶▶▶▶	▶▶▶▶▶▶
Spacecraft		▶	▶▶▶▶▶▶	▶▶▶▶▶▶	▶▶▶▶▶▶
Satellite		▶▶▶	▶▶▶▶▶▶	▶▶▶▶▶▶	
Aircraft	▶▶▶▶▶▶	▶▶▶▶			
Ground-based Systems	▶▶▶▶▶▶	▶▶			
Human Exposure Consequence	Serious illness	Usually fatal	Instant coma		

Figure 3.4 Ranges of total dose exposure for various environments

Neutron Fluence

Neutron fluence (measured in neutrons per centimeter squared) causes permanent damage to semiconductors by upsetting the semiconductor crystal lattice structure. The threshold of effects in semiconductors is $>10^{10}$ neutrons/cm^2.

Single Event Upset

Another radiation-induced effect is known as single-event upset (SEU). Sequential circuits and memories (bistable circuits) are affected when ions from space radiation (high-energy particles) cause a bit to change its state (also known as a soft error). This condition may also lead to latch-up in the device. The use of epitaxial wafers, CMOS SOS or SOI processes and special circuit design techniques provides improved transient radiation hardness and latch-up immunity.

Design for Radiation Environments

All the effects of radiation must be considered when designing a system that may be exposed to radiation environments. The different types of radiation require different hardening solutions. The amount of hardness required depends on the anticipated type and level of exposure and whether the system must simply survive the radiation environment or operate uninterrupted through it. In the latter case, the system must maintain its functionality before, during and after irradiation.

The nature of radiation-hardened processes presents some problems from a design standpoint. Typically, rad-hard processes yield devices that are slower and larger and consume more power than those of conventional bulk processes.

The U.S. government has prepared references for designing rad-hard systems. They include MIL-HDBK-279 and MIL-HDBK-280. The military requirements specifications for rad-hard devices include MIL-STD-883C, Method 5005, Group E and MIL-M-38510F.

Rad Hard or Rad Tolerant?

A rad-hard process is developed to withstand specific radiation levels. A rad-tolerant process, on the other hand, is not designed specifically for radiation hardness, but exhibits some level of inherent resistance to radiation effects. Although rad-tolerant processes are less expensive, their hardness levels may not be guaranteed. Also, radiation-sensitive process parameters, such as oxide thickness, may not be monitored, properly tested, or specified.

When evaluating radiation hardness specifications, look for complete information based on actual characterization and test data. For example, many suppliers will provide the total dose the process will withstand, but will often omit the dose rate and SEU/latch-up tolerance. Understand also what constitutes a device failure. The following should help to determine if any radiation specifications are further qualified:

• Are the specified radiation hardness levels based on gross functional failures or datasheet electrical parameter limits?

• Are the device datasheet parameters specified and guaranteed for post-radiation operation or are derating factors assumed?

• What were the test conditions?

• Were the samples biased or unbiased?

• How soon after irradiation were the devices tested? (This is a particularly important parameter as the device under test can often recover from the radiation-induced effects after a period of time.)

• Was the temperature allowed to rise during the irradiation testing? (Increased temperatures facilitate the annealing or recovery process, producing less than worst-case data.)

Parameter	Bulk CMOS	BiCMOS	Bipolar	GaAs
Speed	Med - High	Med - High	High	High
Power	Low	Low - Med	High	Med
Switching Noise Immunity	Low - Med	High	High	High
Latch-up Immunity	Low - Med	Med	High	1
Production Cost	Low	Med - High	Med - High	High
Integration	High	High	Low - Med	Low
Loaded Output Drive	Low - Med	High	High	Low
Process Complexity	Low	High	Med	Low
Source Availability	High	Low- Med	Med	Low

1. GaAs is a semi-insulating process and, therefore, is inherently resistant to latch-up

Figure 3.5 Comparative summary of ASIC technology alternatives

Chapter 4

Selecting the ASIC Package

THE ASIC PACKAGE is far more than a simple container for the chip. It provides mechanical and environmental protection, as well as the means of mechanical and electrical connection to the system. The package is also responsible for keeping the temperature of the chip within safe operating limits, thus maintaining its reliability and performance.

As the package is often the limiting factor for the design pin count, performance, and cost, it must be treated as an integral part of the design process in order to realize the full benefits of the silicon packaged inside it. The package can be a critical factor in determining the size, cost, and performance of the ASIC, which in turn affects the system size, cost, and performance.

Considerations for package selection include the following:

- Chip size

- Pin count

- Chip heat dissipation (thermal performance)

- Environmental/reliability requirements

- Electrical performance (operating frequency)

- Package material (plastic, ceramic)

- Chip-to-package interconnect (TAB, wirebond)

- Package-to-PCB interconnect (through-hole, surface mount)

- Package footprint (conformance to industry standards)

- In-house or contracted PCB manufacturing capabilities

- Production quantity

- Cost

The relative importance of each criteria for selection depends on the application. As each design element becomes better defined, the appropriate package selection becomes more clear. A great many package and assembly options are available, and the required combination of parameters will likely be satisfied by at least one of them. Take, for example, a large CMOS chip that measures 400 mils/side, requires greater than 200 pins, and dissipates 2 W. Assume also that the chip will operate in a high-volume commercial application with a high degree of price sensitivity. Such an ASIC would likely use either a low-cost, high-pin-count plastic PGA or a plastic quad flat pack, depending on whether the board will be assembled using through-hole or surface-mount technology.

Ideally, the package requirements would be satisfied by using a standard, open-tooled package. However, custom tooling or modification of an existing package tooling is often required. Some of the more common alterations include cavity size shrinks or enlargements and power and ground plane pad reassignments. The design schedule should allow for the time needed for special package tooling, if required. Depending on the type of tooling needed, it may take weeks or months to complete. Keep in mind that test sockets, carriers, and tester load boards may also require special tooling.

As the design progresses, the package specifications will need to be reviewed to ensure that the selected package is still satisfactory. If, for example, the chip size grows beyond initial estimates, it may be too large for the package cavity, or the design may require more pins than originally anticipated.

Package Trade-offs

Packages can be grossly classified by the PCB technology to be used (through-hole or surface-mount) and by their material composition (ceramic or plastic). Package selection begins with decisions based on these two main classes. When coupled with a careful review of the design requirements, the package options can be narrowed considerably. The following overview provides a brief description of the primary package approaches.

Ceramic versus Plastic

Generally, ceramic packages are used in applications where the chip must be hermetically sealed against humid and/or salt atmospheres. Ceramic packages are also far more reliable than plastic for military or other harsh environment applications. Conditions of such environments might include operation at temperature extremes, frequent temperature cycling and vibration. Ceramic, although not a great thermal conductor, is superior to plastic, making it a better choice for high-power chips.

Plastic packages, on the other hand, are suitable for commercial and certain industrial applications where operating environments are nominal. Plastic, however, is a poor thermal conductor. With the larger plastic packages, such as PPGAs, a heat spreader can be molded right into the package. The differing thermal coefficients of expansion between the heat spreader and the package, though, can result in separations where the two materials meet, thus inviting contamination.

In addition to the use of heat spreaders, the thermal resistance of plastic packages can be reduced significantly through the use of copper lead frames. In plastic packages, the lead frame is the key element in heat dissipation. The larger the lead frame, the better it can perform as a heat spreader. Alloy 42, an iron-nickel alloy, has been the most commonly used lead frame material, due to its good mechanical properties. The trend, though, is toward copper lead frames, as they are better electrical conductors and their thermal conductivity is far superior to Alloy 42. Newer copper alloys also have improved mechanical characteristics.

Cost is another differentiator between plastic and ceramic packages. Plastic packages are much less expensive than the ceramic varieties, due to their automated assembly process. In constructing a typical plastic package, continuous rolls of metal lead frames are stamped and cut into strips. The chips are then attached to the lead frames' pads and wirebonded in a highly automated process. The subassembled lead frame strips are then placed in the cavity of a molding press, which is injected with hot epoxy, encapsulating the entire subassembly and leaving only the leads exposed. Hundreds of chip packages can be molded at once in a large mold. The leads are then formed or bent to the desired shape, and the device is tested and assembled onto the PCB. (The newer, finer lead pitches, however, often require that the lead frames be etched, not stamped, which is a more expensive process.)

While plastic packaging has the advantage of low cost, it also has the disadvantage of being permeable to moisture. Thus, the reliability of plastic packages has been lower than those of ceramic. Another reliability problem associated with plastic injection molding is the possibility of voids forming in the encapsulant. A void is missing material or a bubble in the molding compound. Voids have the effect of causing the junction temperature to rise, as well as acting as stress concentrators. Molding compounds, however, are improving and now feature low-viscosity formulas that flow more easily over closely spaced bond wires, thus minimizing the occurrence of voids. Also, as the principal failure mechanisms become better understood, improvements to encapsulation materials have improved their overall reliability.

Differences in the thermal coefficients of expansion between the plastic, the lead frame, the die, and the die-attach material can also cause stress damage to the wirebonds, as well as to the chip. As chip sizes and the number of bonds on finer pitches increase, these problems are exacerbated, particularly under conditions where the package may undergo frequent temperature cycling. New materials and assembly processes, though, are continuing to evolve and promise to minimize moisture permeability and thermal effects and improve plastic package reliability. Packages made with these newer compounds are able to survive a 1,000-hour life test in a test environment of 85°C and 85% relative humidity.

Surface-mounted Packages

ASIC packaging is shifting from through-hole to surface-mount technologies. Because the leads are on all four sides of the package and no holes need to be drilled in the PCB, the packages use less board space than through-hole packages, are less expensive, and can be mounted on both sides of the board. The shorter leads also have reduced inductance, thereby providing better high-frequency performance.

The migration to surface-mount, however, can cause certain problems. For areas where the use of a surface-mount component is a temporary inconvenience, such as in test or board prototyping, surface-mount to through-hole socket adapters can be used.

The engineer faced with a surface-mount design must also consider design for manufacturability. The board assembly requirements must be well understood, as well as how the board will be reworked or repaired in the field. Also, most systems manufacturers don't have the equipment necessary for surface-mount production. There are, however, a large number of contract assemblers who specialize in surface-mount technology.

Fine Lead Pitch Considerations

Advantages offered by fine lead pitches include faster speeds, higher board density, and smaller boards. However, they require special manufacturing capabilities and handling procedures. These devices must be handled in special trays or carriers (which add cost to the component) to prevent damage to fine leads. Alternatively, packages can be supplied with their leads unformed. In this case, the leads radiate horizontally from the package. The leads are formed into the desired shape, typically gull-wing, just prior to being soldered to the PCB. Some vendors also provide a nonconductive tie bar, which, with special carriers, allows the device to be tested prior to lead forming. With the newer tape pack technology, the leads are shipped unformed and are protected by a molded plastic ring or carrier. The package and pins are then stamped out of the carrier so that the leads can be formed.

Finally, the fine lead pitches can present certain problems in manufacturing. The devices are particularly susceptible to solder bridging (which causes shorts between adjacent leads), due to the closeness of the leads. Consequently, PCB design rules for surface-mounted devices require much tighter tolerances than those for through-hole packages. The lands (pads on the PCB where the device is soldered) might be as close as 20 mils, which is much tighter than the 100-mil spacing of conventional through-hole components. Lead planarity (integrity) can also be a problem as the leads are easily bent out of shape. Generally, a maximum of 4-mil coplaner variation can be tolerated, but the amount of solder used to make the connection is also less than that for wider pitches, thus making the solder joint inherently weaker. The J lead configuration is much stronger, but because the leads are bent into the underside of the package, inspection is extremely difficult.

Chip Carriers

Chip carriers are available in a number of materials, shapes, and lead configurations. These include plastic and ceramic, square and rectangle, leaded and leadless. All are intended for surface mounting, although they can also be socketed. Lead counts range from 20 to more than 300.

One of the most popular ASIC packages, due to its small size and low cost, is the plastic leaded chip carrier (PLCC). This package is very practical for ASICs with up to 84 leads. Higher pin counts are available, but such requirements are better served by the plastic quad flat pack (PQFP). The PLCC leads are most commonly formed in the J shape while the PQFPs are generally provided in the gull-wing configuration (Figure 4.1).

Figure 4.1 160-pin PQFP in gullwing lead configuration. (Photograph provided courtesy of LSI Logic Corporation)

53

PQFPs are available in pin counts to the mid 200s. Higher pin counts, to the mid 300s, are available in ceramic quad flat packs (CQFP). These are among the most expensive packages and in many cases cost more than the die. High-pin count CQFPs (Figure 4.2) can cost as much as $0.15 to $0.20 per lead (plus assembly costs), compared to $0.01 total assembled cost (in volume) per lead for PQFPs and PLCCs.

Another chip carrier variety, the leadless chip carrier, is a hermetically sealed ceramic package that has contact pads around its perimeter, rather than pins. Leads are often brazed onto the package in a later operation. When leadless chip carriers are surface mounted without the use of brazed pins, ceramic substrates should be used because the thermal mismatch between a ceramic package and an epoxy-based PCB can cause excessive stress on the solder contacts, which in turn can lead to mechanical failure.

The two competing standards for these and other packages are the US JEDEC and the Japanese EIAJ. The standards call out differing package outline dimensions and lead pitches, among other things. Some plastic packages specify a maximum die thickness, which may require grinding down the back of the wafer to comply with the package profile requirements. The standard to which the selected package complies should be well understood because it can have a significant impact on board design and manufacturing.

Through-hole Mounted Packages

Until recently, through-hole mounted packages were the mainstream packaging technology. They persist due to the wide availability of tooling and manufacturing facility investment, as well as their relative ease of board-level assembly. They are most commonly found in applications at opposite ends of the pin-count spectrum: 12- to 24-pin DIPs in the low-pin count range and 120 to 300+ pin PGAs in the high-pin count range. Although both DIP and PGA packages are available in the intermediate pin count ranges, their use is generally supplanted by surface-mount packages, which offer significant cost advantages, particularly in those middle pin count ranges.

Dual-Inline-Packages

The DIP is the most commonly selected package for pin counts of 40 and under. However, DIPs become very space inefficient at pin counts above 40. The primary

DIP classes are the ceramic DIP (CDIP) and the plastic DIP (PDIP).

Ceramic DIPs come in two varieties: the cerdip and side-brazed DIP. Both are hermetically sealed and provide high reliability. The differences between the two are in the way they are manufactured. The cerdip involves two slabs of ceramic material that sandwich the die-attached and bonded lead frame with a glass (frit) seal. The side-brazed DIP, on the other hand, is a multilayer package that has no preformed lead frame. The leads are attached or brazed onto the sides of the package body. The die is then hermetically sealed in the package with a metal or ceramic lid.

Pin Grid Arrays

Of the high-pin count packages, pin grid arrays (PGA) are the most common. In fact, for through-hole packages with more than 64 leads, it is the only viable option. PGAs are available in ceramic and plastic and in cavity-up and cavity-down configurations. In the cavity-down configuration, the package lid (which covers the cavity opening) is on the same side as the pins. In the cavity-up configuration, the package lid is on the top side of the package (opposite the leaded side).

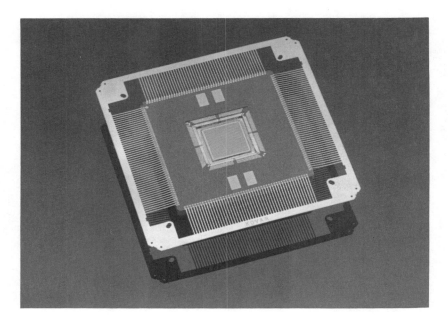

Figure 4.2 Ceramic chip carrier with top-brazed lead frame (prior to lead form and trim). (Photograph provided courtesy of LSI Logic Corporation)

55

The position of the cavity (area where the die is seated), is important in applications where the chip dissipates more than 2 W. In the cavity-down configuration, the back of the die (which dissipates most of the heat) is attached to the surface, which will be exposed to airflow. The airflow across this surface area (which can be enhanced with a heat sink) provides superior heat dissipation characteristics than the cavity-up configuration.

It is important for the designer to understand the bonding pad layout inside the package and the corresponding pin-out assignment. In multilayer packages, power and ground planes are tapped by fixed pad locations. The bonding pads in the chip's periphery must comprehend these locations in the chip layout. Also, the number of package pins that are dedicated to power and ground plane connections must be understood to ensure that the package can accommodate all the signal I/Os. For example, depending on the tooling, a 256-pin CPGA may have as few as 200 pins available for signal I/O, with the remainder being dedicated to power and ground pins.

Figure 4.3 Cavity-up PGA. (Photograph provided courtesy of LSI Logic Corporation)

Figure 4.4 Cavity-down 224-pin PGA. (Photograph provided courtesy of
Motorola Inc., ASIC Division)

Plastic PGAs (PPGA) are a relatively new package technology and, although not as reliable as ceramic PGAs, they are considerably less expensive and exhibit thermal characteristics similar to the ceramic version when an embedded copper heat spreader is used. Also, PPGAs are reported to have better electrical characteristics than the ceramic varieties. This is due to the use of copper traces from the chip to the pins; ceramic packages typically use the higher-resistance titanium-tungsten for the traces.

PPGAs are manufactured in a manner similar to printed circuit boards, and the pins are soldered into plated through holes. The chip is bonded in a cavity that is either recessed or dammed. The cavity is then sealed with either a lid or filled with a resin encapsulant.

Tape-Automated Bonding

Tape automated bonding (TAB) techniques promise to bring ASIC lead counts to 500 or more, but the required equipment is expensive and its availability is still somewhat limited. The TAB process permits a pad pitch (pad width plus minimum spacing between pads) of 4 to 3 mils. By contrast, the mechanical nature of conventional wire bonding requires a pad pitch of no less than 5.5 mils. This difference can have a significant impact on the die size of pad-limited designs. For example, if a pad-limited design has 256 pads, the die size will be approximately 400 mils/side with a pad pitch of 5.5 mils and 300 mils/side with a pad pitch of 4 mils. Thus the 4-mil pitch enables an area reduction of nearly 45%. The use of the smaller pad pitch and TAB assembly also means that pin counts can far exceed the limits imposed by conventional wire bonding methods.

TAB Manufacturing Sequence

Tape automated bonding uses a tape that is much like a miniature flexible printed circuit, supplied in a continuous sprocketed film. It consists of copper foil that is etched on a polyimide dielectric substrate to form the copper lead frame pattern. Each window (also called a frame), which consists of the etched lead frame, is duplicated at regularly spaced intervals on the reel of tape.

The use of TAB requires a special wafer fabrication process called *bumping*. Bumping refers to the addition of metal bumps (typically a gold alloy) to the chip bond pads to provide a platform for bonding. Deposition of the bumps involves an additional mask and processing steps that can add $100 to $200 to the processed wafer cost.

The chip is attached to the tape lead frame which has contacts on the same pad pitch as the bonding pads on the chip (typically 4 mils). These inner leads fan out to contacts on a larger pitch (typically 20 mils) on the outer ring. This wider outer lead pitch is more easily probed and assembled onto the PCB or into a package or some other substrate. Inner lead bonding (ILB) is the process of connecting the tape's inner leads to the chip. ILB can be performed one lead at a time (single-point bonding), one side at a time, or all four sides at a time (gang bonding). Single-point bonding is slower, but it is easier to set up and requires less custom tooling. In fact,some conventional wirebonders can be modified to perform single-point ILB. Following ILB, the chip and ILB can be coated or encapsulated to provide mechanical protection.

When the inner-lead bonding process is completed, the tape is rewound on a reel and sent to the outer-lead bonder. In this step, the TAB subassembly is punched from the tape carrier, aligned to the surface of the PCB or package, and gang bonded via thermal compression or reflow soldering. This process is called outer-lead bonding (OLB) because it is the wider-pitched outer leads of the TAB that are bonded. Typically, OLB is performed by third-party contract assemblers who have made the investment in OLB equipment.

Testing is typically done using slide carriers that hold one tape site and are compatible with standard test sockets. ILB TAB can be provided to the user, or contract assembler, in reel format or in singular carriers. If burn-in is to be performed, they must be provided in carriers.

The reliability of conventional wirebonding diminishes with increased pin count. TAB, on the other hand, is superior to conventional wirebonding in this respect. TAB has greater pull strength; it takes more force to pull and detach a wire from a TAB connection. Where wirebonds may have variations in shape and spacing, TAB leads are shorter, have lower impedance, have a larger cross-sectional area (better thermal properties) and are uniform in shape and spacing. These attributes contribute to lower propagation delays and less signal distortion.

Multilayer TAB tapes (which may include a signal plane, a power plane, and a ground plane, each of which is contacted by leads through vias in the dielectric) exhibit much improved electrical characteristics. Isolating the planes (with dielectric

Figure 4.5 564-lead TAB in carrier. (Photograph provided courtesy of Motorola Inc. ASIC Division)

films between the metal layers) significantly reduces simultaneous switching noise problems commonly encountered with single-layer tapes in high-speed applications.

Because of the high cost of bonding equipment, process steps, and unique tooling required for each design, TAB is only cost effective in high-volume applications. The break-even point compared to conventional wirebonding assembly is at about 100 leads. The costs involved in TAB include wafer bumping, the tape tooling, tooling that may be required to perform ILB and OLB, and the recurring cost for ILB and OLB.

Chip on Board

Chip on board (COB) is a low-cost assembly option for many consumer electronics applications. Since the die is attached and wirebonded directly to the PCB (which may be anything from ceramic to a thin-film hybrid carrier), no lead-frame is required; however, the chip may also be TAB mounted. After wirebonding, the chip is encapsulated with a blob of epoxy, appropriately called a *blob-top*.

COB has a few notable drawbacks: heat dissipation for larger chips can be a problem, and repair or rework is next to impossible. Also, testing is limited to a wafer sort test program. If the chip is found to be nonfunctional after it has been mounted, the board must be discarded.

Multichip Packages

Typically, multichip packages are used to interconnect from as many as 5 to 15 chips. Common configurations include one or more ASICs, a microprocessor, and several memory chips. They may also integrate into a single package diverse technologies such as CMOS and bipolar, as well as digital and analog functions. Multichip modules (MCM), as they are also called, are most often used in applications where interchip delays must be kept to a minimum. The short interconnect lengths minimize chip-to-chip delays and skew and allow for faster operation due to the lower interconnect capacitance.

MCMs can be assembled using hybrid and conventional wire bonding techniques or TAB. They do, however, present certain assembly and test difficulties. Chip-level testing is limited to probe at wafer sort, which may not be very complete or accurate. The packages are very expensive, and if one or more of the assembled chips fail, the entire unit must be discarded. The use of TAB, however, allows more thorough testing (via test points in frame sockets) as well as burn-in. This approach can dramatically improve assembly and final test yields.

Other challenges MCM technologies present include die procurement, test, and quality issues. Depending on the arrangement, the user may be put into a position of having to take ownership of the operations and yields from wafer probe, assembly, test and reliability of the completed MCM. This is due to the inevitable involvement of multiple technologies and vendors required to build the MCM. The ASIC content of the MCM may come from one or more vendors, while the RAM and other components may be supplied by others. The package will be supplied by yet another vendor and other third parties may be required to perform the die prep, assembly, and test operations. In such an arrangement, it is obvious that failures will be difficult to correlate, but fingers easy to point. Multichip module technology may solve many system problems, but one should enter into it with both eyes open so that other problems will not be created.

Figure 4.6 Multichip module package. (Photograph provided courtesy of Unisys Corp.)

Electrical Considerations

As the speeds of ASIC devices increase, the electrical characteristics of the package become more significant. As such, the signal paths between ICs can no longer be treated as simple mechanical connections. Rather, they must be considered as transmission lines where electrical noise is generated by high-speed signals. There are three primary noise sources:

- Crosstalk between adjacent leads

- Reflections resulting from impedance mismatches

- Simultaneous switching of output pins

Crosstalk has the effect of contaminating one signal line with the output of another. This can cause false triggering, background noise, and other signal distortions. Crosstalk is caused by the capacitive coupling between adjacent signal lines and is directly proportional to the signal switching frequency.

Reflections, or output ringing are caused by the impedance mismatch between the output buffer and the transmission line. The transmission line is composed of the bond wire, package trace, package lead, socket and PCB trace. Without an adequate, impedance-controlled driver, problems such as double clocking, false triggering, and supply voltage spikes can occur.

A large number of simultaneously switching outputs can cause voltage spikes on the supply lines and can cause noise-sensitive circuit elements to change their states or clock in the wrong signal. Although these problems have long been characteristic of ECL and GaAs chips, submicron CMOS processes exhibit the same characteristics. Crosstalk can also become a problem with package pin pitches of 10 mils, or less, which has the effect of negating the benefits of finer pin pitches. Although electrical noise can generally be dampened by adding ground pins, reducing the pin pitch does not necessarily increase signal pin density if every third pin must be used for signal isolation. More commonly, ground planes are added inside the package itself. Multilayer packages provide superior electrical characteristics and improved noise immunity.

Capacitance and inductance levels for packages vary widely. The values are of special importance for high-speed applications. Pin capacitance can be as high as 15 pF and inductance as high as 15 nH. The finer lead pitches, though, reduce inductance significantly.

The electrical characteristics of packages can no longer be ignored in the ASIC design phase—particularly in high-speed applications. Rather, their effects must be modeled and simulated. Fortunately, some vendors recognize this and are able to provide SPICE models for their packages' electrical characteristics.

Thermal Considerations

The majority of electronic equipment failures are temperature related. Because the stability of the semiconductor junction declines with increasing temperature, the ASIC design must include consideration of thermal effects on chip reliability. To increase the operating life of the ASIC device, the transistor junction temperatures must be kept to a minimum. Many problems related to heat can be prevented by proper package selection and cooling schemes. Also, in addition to being more reliable, cooler chips perform better.

The parameter with the greatest effect on junction temperature is the thermal resistance between the transistor junction and ambient (surrounding) air. The thermal resistance of a package (the path between the junction and ambient air) is measured in °C/W. This figure relates the temperature rise of the package to the power dissipated by the chip. The primary method of measurement compares the temperature of the transistor junctions, T_j, in the device to the temperature of the surrounding cooling medium, typically ambient air, T_a. This yields the thermal resistance from the junction to ambient air, Θ_{ja}, where Θ is the sum of all the resistances between the two reference points, the transistor junction and ambient air.

The objective of thermal design is to minimize the chip's junction temperature, T_j. The cooler the chip is allowed to operate, the better its performance and reliability will be. Most vendors specify a maximum T_j of 125°C, with 85°C being most desirable. Beyond the maximum temperature, performance and reliability are adversely affected.

The essential strategy for heat removal from the device is a package design that provides a low-resistance thermal path from the chip's junction to the cooling medium. The junction temperature, T_j, is also a function of the power dissipated, P, through the thermal resistance, Θ_{ja}, of the package as follows:

$$T_j - T_a = P\Theta_{ja}$$

This relationship can also be used to show the maximum amount of power the chip can be allowed to dissipate in a given package:

$$P = \frac{T_j - T_a}{\Theta_{ja}}$$

For example, if T_j is to be limited to 125°C, T_a is 70°C and the package Θ_{ja} is 44°C/W (still air), the power dissipated must not exceed 1.25 W. Should the power rise above this limit, the junction temperature will also rise. The options for accommodating the increase in power dissipation include providing airflow over the device and/or adding a heat sink to the package. Either approach will lower the thermal resistance of the package, permitting a greater margin for heat dissipation while maintaining the junction temperature at the desired limit.

The volume of airflow, measured in terms of linear feet per minute (LFPM), has a significant effect on the package Θ_{ja}. An air flow of 500 LFPM can nearly halve the package Θ_{ja} in the still air condition.

Certain package materials, such as BeO, AIN and SiC, exhibit very high thermal conductivity. When coupled with a copper lead frame, an external heat sink and a large volume of airflow, a Θ_{ja} as low as 2°C/W can be attained, permitting power dissipation of as much as 30 W.

Chip size also has an effect on the package Θ_{ja}. Silicon is an excellent thermal conductor, so the larger the chip area, the better the Θ_{ja}. For this reason, silicon substrates in packaging are beginning to see increased use. Another advantage to the silicon substrate is that there is no thermal coefficient of expansion mismatch, thus limiting stress at die attach and wirebond joints.

For ECL chips dissipating more than 30 W, liquid cooling may be required to keep junction temperatures within safe operating ranges. For devices dissipating less power, a package with a heat sink and several hundred linear feet per minute of forced air cooling is generally adequate.

Package Type	Mechanical Characteristics	Electrical Characteristics	Thermal Characteristics[1]	Package Cost	Package Reliability
PDIP	Pins: 8 to 64 Pin Pitch: 100 mils	R: Med L: High C: Low	Fair	Low	Good
CDIP	Pins: 8 to 64 Pin Pitch: 100 mils	R: Med L: High C: Med	Good	Low	High
PLCC	Pins: 28 to 84 Pin Pitch: 50 mils	R: Med L: Med C: Low	Good	Low	Good
PQFP	Pins: 48 to 224 Pin Pitch: 10 to 25 mils	R: Med L: Med C: Low	Fair	Low	Good
CQFP	Pins: 48 to 340 Pin Pitch: 10 to 25 mils	R: Med L: Med C: Med	Good	High	High
CLCC	Pins: 20 to 84 Pin Pitch: 40 or 50 mils	R: Med L: Med C: Med	Good	Med	Very High
PPGA	Pins: 48 to 256 Pin Pitch: 100 mils	R: Low L: Low C: Low	Good	Med	Fair
CPGA	Pins: 48 to 390 Pin Pitch: 70 or 100 mils	R: Low L: Low C: High	Very Good	High	Very High
TAB	Pins: Up to 500 Pin Pitch: 10 mils (OLB)	R: Low L: Low C: Low	Depends on substrate and encapsulation	Low	[2]

R = resistance, L = inductance, C = capacitance.
1. Assumes still air conditions. Can be improved with airflow, heat sink, copper lead frame
2. TAB is more reliable than wirebonding. Overall reliability depends on substrate/encapsulation.

Figure 4.7 Comparative summary of ASIC package alternatives.

66

Chapter 5

Selecting the ASIC Design Tools

THE SELECTED ASIC DESIGN TOOLS will have a tremendous impact on the quality of the design, as well as the ease of the design process itself. A rather diverse suite of tools is required to completely execute an ASIC design. These include tools for design entry, simulation, layout, verification and design documentation.

Proper design tool selection is one of the keys to minimizing the risks of design delays and iteration. Software bugs and kludgey interfaces can severely affect productivity and design schedules. Also, incorrect behavioral models or poorly calibrated timing models can render a circuit inoperable. Keep in mind, though, that, although the quality and configuration of the tools are critical, even the best design tools in the hands of a poor designer will be no match for a good designer equipped with adequate tools.

In this chapter, each constituent element in an ASIC design environment will be reviewed. Specific strategies will be presented for evaluating, acquiring, and integrating the most appropriate tools into an optimal design environment—based on the needs of the users and the specific requirements of the design project(s). The *bringing up* of an ASIC design environment involves a significant long-term resource commitment, therefore, decisions in this area should consider both current and future needs. They should also comprehend the organization's ability to staff and maintain the environment. With time, user expectations will inevitably rise and needs will outgrow the available resources. Obsolescence is another fact of life. The message in all of this is that a *continuing* investment will be required to maintain an adequate in-house ASIC design capability. Change, it seems, is the only constant. When a preferred ASIC vendor no longer supports your preferred choice of platform, or the design tool vendor goes out of business or is unable to support new technologies, one must be able to respond. To protect the large CAE tool investment, the design environment must be *flexible* enough to accommodate such changes.

The CAE tool investment involves far more than just the cost of the hardware and software. The time required to fully implement a design environment may be a year or more. In addition, many months of hands-on training and experience are needed for users to gain expertise with the tools. Bringing new engineers into the system will also require time and money. Perhaps the best strategy for acquiring, installing, and assimilating new CAE tools is to develop the design environment in stages—integrating only a few applications at a time, and then adding tools as they are needed. Taking one bite at a time will also relieve much of the organizational stress, as acquiring these tools can be expensive, risky and difficult.

Schematic Capture

PC-based schematic capture packages are the most popular method of entering ASIC designs. Although many vendors bundle their schematic editors with other workstation and PC-based design automation tools, most are available as stand-alone products. The market for these packages is extremely competitive and though many PC-based packages are available for well under $1,000, there can be many subtle differences between them.

The primary areas of evaluation should include the user interface, schematic editing flexibility and ease of use, interface to other design tools, and third-party library support. The evaluation should include a checklist of considerations and specific desirable features, which might include the following:

User Interface

- Command entry flexibility (including command line, customizable pull-down and pop-up menus, function keys, and mouse buttons).

- Windowing capability (allows simultaneous viewing of multiple schematic sheets).

Schematic Editing

- Number of levels of design hierarchy supported.

- Ease of moving between different levels of design hierarchy.

- Ability to cross-reference, place, locate and check design rules across multiple schematic sheets (typically a function of the ability to maintain the entire design database in RAM).

- Ability to extend nets across sheets.

- Ability to specify critical path placement and routing.

- Instance manipulation (including rotate, mirror, and so on).

- Ability to drag wires (with and without components attached) and cells (with ghost-imaging for more accurate placement).

- Symbol editor ease of use.

- On-line design rules checker (flags opens, shorts, excessive fanout).

- Number of zoom levels supported.

- Panning capability.

- Variety of text fonts and sizes and ease of attachment.

- Availability of Undo and Undelete commands, as well as the ability to save deletes in a separate file for later Undos, if needed.

- Support of repetitive operations.

- Rubber-banding of wires and buses, snap to grid, snap to pin.

- Reports, including an unwired pin list, netlist, cell list, and so on.

- Sheet capacity/memory requirements.

Interface/Integration with Other Design Tools

- Support of standard data formats, including EDIF (electronic design interchange format).

- Ease of integration to other tools.

- Interfaces for simulators, timing analyzers, and physical design tools.

- Ability to view schematic and simulator output on same screen.

- Ability to display simulation waveforms on the schematic at node of interest.

- Netlist formats supported.

- Printers and plotters supported.

Third-party Library Support

- Is the selected ASIC vendor's cell library available and qualified for the schematic editor?

Logic Synthesis

Design capture via logic synthesis involves specifying designs using logic equations, state transition diagrams, finite state-machine languages, and truth tables. It is a highly leveraged design approach because the emphasis is on the description of the circuit behavior, rather than the specific gate-level implementation. This ability dramatically improves productivity, as well as design efficiency for both combinational and sequential logic blocks.

Logic synthesis performs automatic translation of a high-level design description into a specific, optimized logic implementation. The designer is able to obtain fast, accurate feedback on the synthesized block's speed, size, and power consumption and perform comparisons of a variety of circuit implementations. But the main advantage synthesis offers is that the design of complex control and decoding functions can be completed in a fraction of the time over manual, gate-level design methods.

Logic synthesis tools begin by minimizing the entered logic equations by identifying common logic terms. The minimized equations are then mapped into a physical architecture. Here, the user has several options. The design may be mapped into gates, PLA structures, complete with internal registers (similar to those of programmable logic), decoders, or combinations of these elements (Figure 5.1). The most flexible tools allow the designer to target his design to his choice of programmable logic, gate arrays, and standard cells.

In the case of the gate-level implementation, the user must first specify the ASIC library that is to be used. The synthesis tools then map the design into a netlist of the logic gates and elements available in that library. Synthesis output formats include netlist, schematic, JEDEC fuse maps (for PLD programming), as well as compiled physical layout (through links to physical design tools).

The use of synthesis tools, as can be inferred, provides a degree of technology and vendor independence (at least for the parts of the design that can be described and efficiently implemented using these tools). Generally, though, logic synthesis is used in conjunction with other tools, such as module compilers and block place and route tools, which may be supplier specific. The blocks created with synthesis tools (usually control logic), along with other compiled functions (such as RAM, multipliers, and other datapath elements), are then wired together using a conventional schematic capture program. For this reason, the synthesis program should automatically generate a symbol for use in schematic entry.

Later, when the synthesized blocks are combined with other elements during layout, there may be an option for flattening them (breaking them down into primitive gate elements from higher-level functions, such as flip-flops) so that placing and routing can be based on the most efficient floor plan for the ASIC. Since the blocks' constituent elements are effectively dispersed among the other blocks in the ASIC, the timing will affected. The circuit must, therefore, be resimulated to take into account the new routing delays. Otherwise, the synthesized logic can be kept together and placed as a block, keeping its timing model intact.

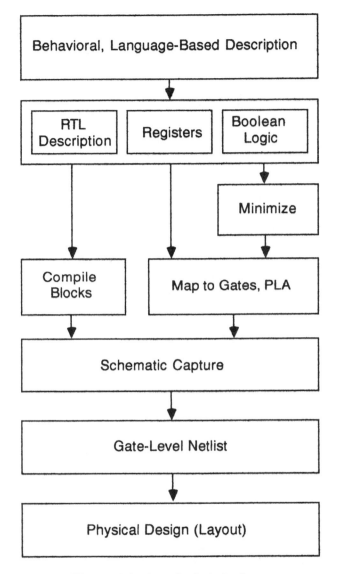

Figure 5.1 Logic synthesis design flow.

In addition to behavioral and gate-level descriptions, there is an intermediate level known as the register-transfer level (RTL), that is also generally supported by synthesis tools. The RTL description is more structural in nature, as it describes the data flow from register to register. It also distinguishes the data paths from the control paths.

Another important function of synthesis is logic minimization, or optimization. Logic minimization is used to eliminate redundant portions of logic and can improve speed, area, and testability of the design. It may also be performed under a wide range of optional user-defined constraints, such as drive and timing, as well as areas that are not to be optimized.

Evaluating Logic Synthesis Tools

Writing a design description in a logic synthesis environment is much like programming in a high-level language, such as C or Pascal. VHDL (the DoD-mandated VHSIC Hardware Description Language), though, is expected to become the primary synthesis language. In the meantime, proprietary languages abound.

Learning, using and maintaining multiple proprietary languages in a design environment can become extremely difficult. Therefore, the use of an industry-standard language, like VHDL, should be an important consideration in selecting synthesis tools.

When evaluating third-party logic synthesis tools, the most important considerations are the results they produce. What is the speed and area of a block in a given technology? Will the synthesized block be testable? Rather than relying on vendor claims and benchmarks, the user should prepare and perform his own benchmarks that will be representative of designs to be done.

Other important considerations include compatibility with existing design tools, methodologies, and ASIC libraries. Also, since the majority of ASIC design starts are prompted by the redesign of an existing system, and since most existing designs are described in a netlist, not a language, the synthesis tools should be able to operate on designs in existing, foreign netlist formats.

Modeling Behavior

As designs become more complex, gate-level descriptions become unmanageable, making it necessary to describe large designs in a higher-level, more abstract manner.

73

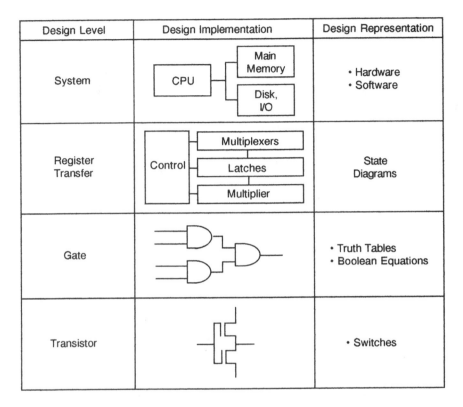

Design Level	Design Implementation	Design Representation
System	CPU — Main Memory / Disk, I/O	• Hardware • Software
Register Transfer	Control — Multiplexers / Latches / Multiplier	State Diagrams
Gate		• Truth Tables • Boolean Equations
Transistor		• Switches

Figure 5.2 Design representation views.

Behavioral models, developed using hardware description languages (HDLs), provide at least a partial solution. By using behavioral models, simulation can be performed at a very early stage of development—even before a structural implementation exists. This approach facilitates and even encourages the exploration of architectural options.

HDLs are extremely versatile, and are finding use not only as a means of describing hardware, but also as a means of providing behavioral modeling for system-level simulation and for logic synthesis. Where the HDL is an industry standard (such as IEEE VHDL), it provides a means of transportable design documentation.

Although HDLs can be used to describe entire systems, they cannot represent a specific physical layout or even a schematic. The behavioral description defines what the block does, not how it does it. To realize a physical implementation of a design, the HDL must be integrated with cell libraries (or module compilers), synthesis, schematic capture, simulation, and layout tools.

The high-level behavioral description must be mapped into a specific low-level description that can be operated upon by physical design tools. To completely verify a design, gate-level simulation, back-annotation of wiring delays and timing analysis are still required.

One key advantage of an HDL design environment is that the design is developed in stages. This greatly simplifies simulation, because as each block becomes more defined structurally, it is easier to isolate and debug.

In a typical design flow, the designer begins by writing a behavioral model that describes the entire system. This high-level model is then partitioned into individual functional blocks, some of which will be ASICs; others will be standard components. At this stage, each block is treated as a black box, with inputs, outputs, and a specified delay between them.

Once the behavior of the ASIC blocks are specified, the design can be developed at a more detailed, register-transfer level. At this point, specific logic building blocks (compiled, synthesized, or selected from a cell library) can be used to implement the required behavior (via conventional schematic capture) until, ultimately, a gate-level structure results.

When the ASIC design has been completely decomposed to the gate level, a simulation of the complete system can be performed, using gate-level models for the ASICs and behavioral models for the standard components. (The use of both gate- and behavioral-level models in a single simulation is referred to as mixed-mode simulation). Finally, the results of this mixed-mode simulation are compared with the original top-level behavioral model to ensure that the lower-level functions, working together, match the top-level, behavioral-level simulation.

Behavioral models for standard components (ranging from SSI parts to RAMs and microprocessors) can be purchased from simulation vendors and third-party modeling vendors. These models are generally very accurate and are verified against the actual part. Full-function models, that are capable of executing machine code and displaying internal register states, are available for many microprocessors. These models are essential for simulating ASICs which will operate in microprocessor-based systems. Another advantage of using behavioral modeling is that it simulates much faster than a gate-level model of the same function.

Complex standard parts, such as microprocessors and peripheral controllers, cannot be modeled at the gate level because schematics for proprietary devices are generally not available. Besides, they would be extremely difficult to develop and would also increase simulation times. The information provided in a device datasheet, though, is generally sufficient to model the behavior of a device. The timing information provided in the datasheet also reflects guardbanding to account for lot-to-lot variations, as well as the range of operating conditions.

VHDL

There are a great many HDLs on the market, each with its own proprietary characteristics. But VHDL (per IEEE-1076) is emerging as the industry standard. Some vendors claim to support the entire VHDL standard, while others support only a subset. Others, support VHDL through translation.

The theory behind the Department of Defense's mandate for a standard design documentation format (VHDL) is that it will allow them to take a VHDL design specification to any number of ASIC vendors and have them produce a functionally equivalent chip. Ultimately, a complete VHDL design environment will allow behavioral to structural synthesis for all circuit functions (not just control logic).

The success of VHDL, then, is heavily dependent upon achieving the support of ASIC and CAE tool vendors. Initially, to get the ball rolling, vendors have provided translators to and from VHDL descriptions for their simulators. Translation, though, presents several problems. First, the limits of the host simulator necessarily impose limits on the extent of VHDL that can be accepted. In other words, a translator will provide only a subset of the VHDL capability. Second, debugging can be difficult in the translated version because the design was developed in a different environment. Finally, keeping things straight in *glued* design environments can become very confusing for the user. Fortunately, CAE vendors are beginning to respond by providing integrated VHDL design environments.

Today, VHDL is most commonly used to satisfy the DoD documentation requirements. Postprocessors are able to generate a VHDL description from a netlist created using conventional tools. It is anticipated, though, that the DoD will also require that designs be simulated in VHDL and that test vectors in VHDL format be provided as well.

Moore Machine

This example illustrates a basic Moore
finite-state machine, summarized in the state
diagram and state table below. The VHDL
code to implement this finite-state machine
follows. Note that the machine is described
in two processes: one declares the synchronous elements of the design (registers); the
other declares the combinational part of the
design.

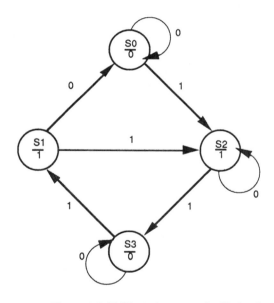

Present state	Next state		output (Z)
	x=0	x=1	
S0	S0	S2	0
S1	S0	S2	1
S2	S2	S3	1
S3	S3	S1	0

Figure 5.3 VHDL design example. (Illustration provided courtesy of Synopsis, Inc.)

```
- - Moore machine                                              else
entity MOORE is                                                 NEXT_STATE <= S2;
  port(X, CLOCK: in BIT;                                       end if;
      Z: out BIT);                                             when S1 =>
end;                                                            Z <= '1';
                                                               if X = '0' then
architecture BEHAVIOR of MOORE is                               NEXT_STATE <= S0;
  type STATE_TYPE is (S0, S1, S2, S3);                         else
  signal CURRENT_STATE, NEXT_STATE: STATE_TYPE;                 NEXT_STATE <= S2;
begin                                                          end if;
                                                               when S2 =>
  - - Process to hold synchronous elements (flip-flops)         Z <= '1';
  SYNCH: process                                               if X = '0' then
  begin                                                          NEXT_STATE <= S2;
    wait until CLOCK'event and CLOCK = '1';                    else
    CURRENT_STATE <= NEXT_STATE;                                NEXT_STATE <= S3;
  end process;                                                 end if;
                                                               when S3 =>
  - - process to hold combinational logic.                      Z <= '0';
  COMBIN: process(CURRENT_STATE, X)                            if X = '0' then
  begin                                                          NEXT_STATE <= S3;
    case CURRENT_STATE is                                      else
      when S0 =>                                                NEXT_STATE <= S1;
        Z <= '0';                                              end if;
        if X = '0' then                                      end case;
          NEXT_STATE <= S0;                                  end process;
                                                           end BEHAVIOR;
```

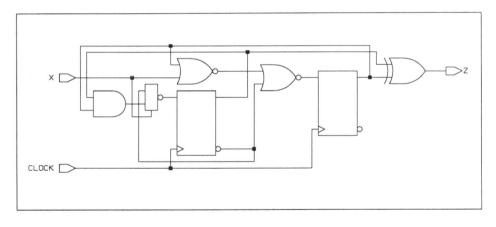

Figure 5.3

Many engineers are reluctant to adopt VHDL as a design methodology because it involves learning to write code and to express circuit behavior in a language as opposed to a schematic (Figure 5.3). However, once they do, they will be able to execute designs better and faster.

Simulation

Far too many ASIC designers have discovered the hard way that whatever doesn't get simulated usually doesn't work. Rework is extremely expensive, and the time it requires may push the product introduction outside its market window. Time to market pressures are compounded by the amount of time it takes to perform a thorough simulation. Nonetheless, simulation is the key to assuring that the end product will work according to specification.

The ASIC designer must not only simulate the ASIC's function, but also the ASIC's behavior and interaction within the system, as well as the effects of temperature and supply voltage variation. The process in which the ASIC will be fabricated will also vary from fabrication lot to lot. The combined effects of temperature, voltage, and process variation can increase nominal timing by more than twice. Since the variations affect the chip globally (the delays tend to track), the part may be all fast or all slow, depending on where the lot falls in the distribution.

Modeling these effects in simulation is statistical in nature. Derating factors are based on sample data and process control monitors. The simulator must be able to vary these parameters to determine minimum, maximum, and nominal delays. Overoptimism with these variables can result in chips that don't work. Excessive conservatism, on the other hand, may keep the engineer from fully exploiting the capability of the process, thus limiting the design's competitiveness. The key is a complete understanding of the system and its operating conditions and environments. Since process distribution data are generally considered to be sensitive, proprietary information, vendors will typically release only derating factors. Still, the ASIC may exhibit performance that will vary depending on how tight the process controls are and how the process was running at the time the lot was processed. There can be significant variations in lot-to-lot performance. It is not uncommon, for example, for an ASIC that is specified and tested to operate at 40 MHz to actually operate at 50 MHz. Many a foolish engineer has assumed that the vendor has guardbanded the specification unreasonably and, therefore, takes the liberty to increase the system clock frequency. The folly becomes painfully evident when the next lot is marginal at 40 MHz.

Simulation Overview

Simulation is generally performed at some combination of behavioral, structural, and hardware levels. The different model types are complementary and, when used together, provide an optimal simulation environment. An essential criteria for selecting a simulator, therefore, is that it allow all three to be used in the same simulation.

Logic simulators fall into two basic categories: functional and timing. Functional simulation is used to verify that the design behaves as expected, independent of timing delays. Functional simulations greatly simplify timing by assuming that all logic elements have a common delay of one time unit (unit delay). Since delays due to fan-out loading and wiring capacitance are not comprehended, the simulations run much faster, and the designer is able to obtain quick feedback on circuit functionality.

Conversely, timing simulation provides detailed information on both the functional and timing aspects of the design. The accuracy of timing simulation is heavily dependent on the accuracy of the models, as well as its determination of delays due to fan-out, net length, and input slew rate. The actual delay values for these conditions are known only after the layout is completed; however, timing simulation can be performed prior to layout by using statistical data that model the effects of layout. Pre-layout timing simulation is useful in locating and correcting gross timing problems prior to layout. Once the layout is completed, the actual delays are *back-annotated* to the simulation so that a more precise evaluation can be performed.

In addition to circuit timing information, the timing simulation also checks for and flags timing violations (setup and hold and minimum clock-pulse width violations, for example) and glitches. Glitches result from input transitions, or voltage spikes, which occur in a time period that is shorter than the delay through the logic element. Glitches in turn can result in erroneous outputs. Since only the engineer can accurately determine whether the glitch will really cause a problem, the simulator should issue a hazard report detailing the node and signals involved.

Simulators are available in three basic forms. They include software simulators, hardware-accelerated simulators, and hardware simulators. The most common

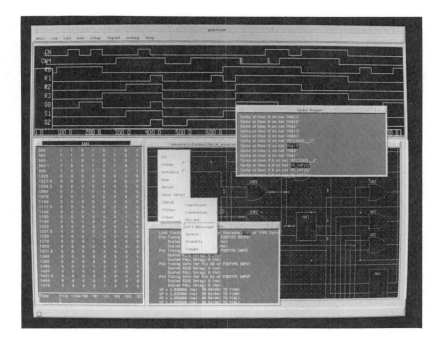

Figure 5.4 Windows showing simulation waveforms, vectors, spike warnings and schematic. (Photograph courtesy of Mentor Graphics Corp.)

implementation is the software form. Software simulators range from simple, PC-based functional simulators to complex timing simulators that run on mainframes.

Although the efficiency and speed of software simulators is improving (as a result of better algorithms and more compact data models), the speed of software simulators is inversely proportional to the size of the circuit being simulated. In fact, for large designs (greater than 20,000 gates), full timing simulations can take many hours or even days to run.

Hardware accelerators relieve much of the simulation bottleneck by speeding up the software simulation algorithm in special-purpose hardware. The accelerator relies on the host software simulator for the user interface and for the libraries. To couple the accelerator to the software simulator, special interface software must be provided or developed.

The long simulation times associated with software simulators can often discourage iteration. Accelerated simulation, because it is fast, allows the engineer to iterate the design until it is optimized.

The following should be considered when evaluating an accelerator:

- It should run the selected simulation software.

- It should support the selected ASIC vendor's libraries (functional and timing).

- It should support the addition of new library elements quickly and easily.

- It should support the types of simulations to be performed, such as logic, fault, timing and behavioral modeling.

- It should have the capacity (expressed in primitives) to run the most complex designs. Gate capacity beyond the base configuration is typically increased by purchasing add-in processor cards.

- It should run large simulations reasonably fast. The throughput time should also include the time required to compile (translate the netlist into the primitive elements the accelerator can use) the design for simulation.

- It should be able to operate as a network resource.

Finally, hardware simulators are stand-alone systems that execute the simulation algorithm in hardware. These systems combine the best of both worlds.

Simulation Vector Generation

To run a simulation, a set of input stimuli must be prepared. These consist of groups of logic 1s and 0s that are applied to the circuit inputs. The results of the circuit operation on these inputs are then compared to a corresponding set of expected output results. If they match, the circuit passes. If not, the bits in violation are highlighted and the debugging begins.

Developing input stimuli, or vectors, can be done manually (the engineer enters each vector individually) or graphically (the engineer draws stimulus waveforms in a graphics environment). The simulation output can be provided in a table of 1s and 0s, which show the state of signals at each time step, or in a waveform format. The waveform output format is generally more intuitive because relationships between data and timing are graphically illustrated.

If the design will include macrocells, the vendor will typically have a prewritten test program for the macro function(s). This relieves much of the burden for the designer, but a global test program will still be required to test the entire circuit.

Simulators can be run in batch or interactive modes. Interactive simulation has the advantage of allowing the user to focus on specific areas of the design, which eases debuggging. The simulator's ability to assist in the debugging of the design is its real strength.

Automatic test vector generation produces random patterns for input stimuli and records the resulting corresponding outputs. Its use should be limited to increasing the level of fault coverage (detection of manufacturing defects), and not for debugging and verifying proper design functionality. If the design is wrong, the randomly generated vectors will never show it. It will always simulate correctly.

ASIC Cell Library Considerations

The *design* information is passed to the simulator via a netlist (that describes the connectivity of all the elements or cells in the design). The *simulation* information is stored in a separate file called a library. The library consists of primitive logic modeling elements (gates) that form the basis for building models for each element in the design. Every cell to be simulated must ultimately be described in terms of these primitive elements.

The model comprehends the functional or logical behavior of the element, as well as its intrinsic and layout-dependent delays. The library is maintained in a separate, external file because each ASIC vendor's library is unique. Also, the models may need to be translated to be compatible with the target simulator. Because primitives are simulator specific (each has a different format and method of recording delay information), ASIC and simulator vendors alike have considerable difficulty supporting each other, and the development costs for each additional simulator or library to be supported are substantial. Maintenance costs are also significant. The development of model timing parameters for a given ASIC cell library is not a one-time occurrence. As process parameters change and new cells are added to the library, the models need to be updated.

Because of these difficulties and the inevitable errors that result, some ASIC vendors will only guarantee the accuracy of simulations run on their *golden simulator*. This is the simulator by which their library is characterized and qualified. Although the ASIC vendor may support other simulators, it is important to understand to what extent the models have been qualified and are guaranteed.

Debugging the Design

Debugging loops can have the greatest impact on overall design development time. Debugging involves isolating the problem and modifying the design and/or the vectors. Typically, the designer will first implement the design using a schematic capture program and generate a netlist. Then, when the netlist is simulated, if errors are discovered, the designer must exit the simulation environment to make the necessary changes in the schematic. The time it takes to go through the steps depends on the time needed to edit the schematic, extract the new netlist, restart the simulator and simulate. Some simulators require that the netlist be expanded or flattened to the simulation primitive (gate) level, removing all design hierarchy. This step alone, depending on the size of the design, may take hours to complete.

Simulating Designs in Submicron Processes

Submicron process technologies present a whole new set of problems for simulators. As system clock rates increase, secondary effects that could safely be ignored by simulators at lower frequencies become significant. In submicron processes, the layout-dependent effects are more critical than the intrinsic gate delays in determining timing characteristics. In fact, interconnect delay may constitute as much as 70% of the total delay of a given path. The narrower wires have higher resistance, and the parasitic capacitance represents a greater proportion of the total circuit capacitance. As a result, for submicron technologies, it is not appropriate to lump RC delay factors into a single value attached to a single node. Rather, the RC must be distributed along the length of the net. Of course, such analysis can only be performed if the actual net lengths are known. Statistical measures of net capacitance, therefore, become less useful when simulating designs in submicron processes.

With the overall faster switching times of gates, the rise and fall delays must also be modeled more accurately. Many delays are cell instance dependent. That is, their delay depends on how they are used in the circuit. For example, the rise and fall times for a cell that is heavily loaded will be much slower than one that drives only one other cell.

Timing Analysis

Timing analysis is performed to detect timing problems that may otherwise go undetected in the logic simulation. Timing errors, either gross or marginal, can show up in the form of nonfunctional devices, lower production yields or reduced product reliability. Therefore, it is critical that the logic simulation be complemented with a thorough timing analysis.

Timing analysis can be either static or dynamic. Dynamic timing analysis evaluates circuit timing while applying simulation vectors at speed. Static timing analysis, on the other hand, compares propagation delays to specified timing constraints. Dynamic timing analyzers require test pattern or vector generation. Because the dynamic timing analyzer is pattern dependent, it can only report errors that result from those patterns. If the patterns do not exercise an error condition, potential timing errors may go undetected.

Static timing analyzers examine circuit timing by adding up propagation delays along paths between clocked elements. Because they are path oriented, they can determine and report comprehensive path statistics, such as the total number of paths, delays between pins in a path, and the longest (or slowest) and shortest (fastest) paths.

Static timing analysis is stimulus pattern independent and, therefore, does not consider circuit behavior in its analysis. As a result, the static timing analyzer may trace paths that are logically impossible or it may report an erroneous worst-case or critical path. The timing analyzer should, therefore, provide a path-pruning capability that allows the engineer to disable such paths in the analysis. In addition, the following capabilities should be considered when evaluating timing analyzers:

- Identify and report setup and hold violations.

- Identify and report path delays exceeding clock periods.

- Calculate the delay between user-specified nodes.

- Allow multiple clocks with different periods.

- Incorporate interconnect resistance as well as capacitance in timing analysis.

- Allow buffer-sizing trade-offs to be performed.

- Analyze circuit descriptions employing mixed models.

- Graphically display the critical paths on the schematic.

- Use the back-annotated delay information in timing analysis.

- Identify areas where circuit performance can be improved.

- Detect paths that are too fast (race conditions) or too slow.

- Include clock skew in delay calculations.

Hardware Modeling

As previously discussed, the use of models for standard components in conjunction with a model of the ASIC (behavioral or gate level) in system simulation is indispensable. But in cases where a behavioral model of a needed standard component is not available (as is often the case for new microprocessors), hardware modeling may provide the answer. A hardware modeler may also be used to model the ASIC once silicon is obtained so that a complete system-level simulation can be performed.

Hardware modeling differs from behavioral modeling in that it uses the actual device to model its own behavior. In many ways, the hardware modeler works like a functional tester. The element to be modeled (which may even be an entire printed circuit board) is mounted on an adapter board and connected to active pin electronics. The modeler then takes the input stimulus from the simulator and returns the actual responses (timing information included).

Typically, the user must first write a device specification file for the modeler. This file includes the device pin-out, defining the I/O and power and ground pins, clock frequency, and other timing information derived from the device data sheet.

ASIC Emulation

The quality of a design verification is a function of the quality of the simulation vectors. The vectors must assure that the ASIC will operate properly in the system under all system conditions and modes of operation. Because breadboarding a complex ASIC design is not practical, the development of a comprehensive simulation places a tremendous burden on the designer.

A special type of hardware modeler, called an ASIC emulator, provides a means of automatically breadboarding the design in hardware. When using an ASIC emulator, the logic is partitioned across and down-loaded to an array of reprogrammable logic devices. An automatic router, based on a matrix of switchable interconnects, completes the connections between devices. The down-loaded design can then be plugged into the target system board's ASIC socket (via cable or pod) and run. This arrangement offers a degree of real-time operation that would otherwise take a tremendous amount of time to simulate.

ASIC emulation, however, sacrifices design timing accuracy because the reprogrammable logic devices have their own timing characteristics. Its usefulness, therefore, may be limited to low-speed (5 MHz, or less) or functional emulation and debugging only. Because of the performance limitations and timing inaccuracies, simulation and timing analysis are still required. ASIC emulation should, therefore, be considered a complementary process in the design cycle. Compatibility with desired libraries and design tools should be assured.

In spite of its limitations, the use of emulation can be extremely valuable in resolving design errors, ambiguities, and specification miscommunications without having to wait for silicon and incurring the risk of iteration. The system also allows software developers to test their code as soon as the ASIC design can be down loaded.

Physical Design

The traditional method of ASIC design has involved the user completing a design netlist, using logic building blocks from the ASIC vendor's cell library, and

87

performing a simulation based on the vendor's timing models. The simulated netlist is then passed on to the ASIC vendor, who completes the physical design. That is, the cells in the netlist are placed and routed according to their specified connectivity. The actual routing wire lengths and their associated parasitic capacitances are then back-annotated to the simulation files. The design database is returned to the user, who resimulates the design to ensure that the post-layout timing effects haven't introduced any timing problems. At this point, the design is ready for final verification and fabrication.

This conventional design flow has several undesirable shortcomings. For example, the user must rely on statistical wiring delays when performing simulation. Obtaining actual layout-dependent delays is an iterative process, requiring submission of the design to the vendor for layout each time the design is changed. The user also foregoes control in the layout portion of the design cycle.

Today's more powerful workstations, though, permit physical design (placement and routing) to be performed by the user at his own location. The type of physical design tools needed depends on the type of ASICs designed and the extent to which the user wishes to become involved with the details of the layout.

The basic layout methodologies include row-based standard cell (pitch-matched height), rectilinear block (random sizes and shapes), analog, and mixed analog/digital. Certain tools may be optimized to work with only one type, while others are able to mix them in the same layout. The technology used (CMOS, Bipolar, BiCMOS, GaAs) will also affect design tool selection, as the different processes all have different layout requirements. Also, support of process capabilities, such as three- and four-layer metal interconnect routing and the ability to perform over-the-cell routing will be key considerations. If working with gate arrays, the selected or preferred ASIC vendor's libraries must be supported. The layout tools must know about the specific array topology, as well as its routing resources.

If in-house layout verification is a requirement, links to layout verification tools (LVS, ERC, and DRC) and a graphics-based geometry editor (for correcting layout errors) will be needed. These links require interfaces for standard geometry input and output formats such as GDSII or CIF, which may be offered as optional utilities by the tools vendor.

Floor Planning

Floor planning is a process of placing functional blocks within the chip area and allocating the routing spaces between them such that an optimum layout is achieved. Floor planning can be automatic or interactive. Manual placement allows the designer the freedom to preplace cells constituting critical paths, as well as to make subtle design improvements.

Blocks that communicate heavily with one another (via common nets) should be placed close together to minimize routing distances and congestion. Reducing congestion increases the probability of 100% completed autorouting. Net lengths also need to be kept as short as possible so that timing will not be adversely affected. By specifying the critical nets, the router will know specifically which nets need to be short and which can be allowed to meander to reduce routing congestion.

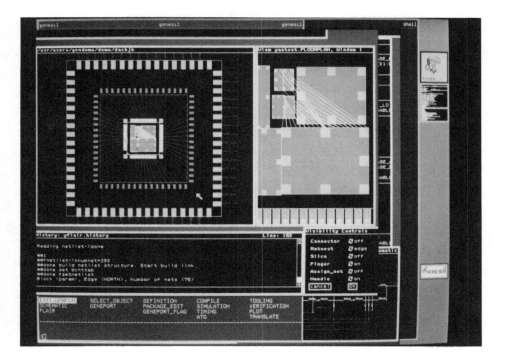

Figure 5.5 Physical design windows showing floor planner (with force vectors turned on) and packager. (Photograph courtesy of Mentor Graphics Corp.)

Routing congestion in cell-based designs can cause the die size to grow. In array-based designs (where layout must work within the constraints of a fixed die size), the area required due to routing congestion can exceed the capacity of the array, forcing the user to go to the next higher array size.

Layout Optimization

The extent of layout optimization depends greatly on the array architecture, the number and types of cells used, and their connectivity. Typically, the routing-to-active area ratio is 1:1, however, less structured designs or designs with many large buses or many feedback loops may boost the ratio to 2:1 or more. In addition to design-specific concerns, the efficiency of the layout is also a function of the quality of the layout tools. Once again, a benchmark will reveal the differences in efficiency between tools.

Dramatic differences in results can often be seen between various methodologies and tools. The same design in the same process designed with different libraries and layout tools may have an area delta of 2:1 between them. Some layout methodologies require the use of compaction tools to improve area efficiency. Compaction programs minimize the spaces between blocks and nets to ensure that all layout elements are as close together as the process design rules will allow. Compaction will not only reduce die area, but may optimize performance as well.

As chip sizes grow, clocked elements, such as registers, become widely distributed and are connected to clock signals through long signal paths. The skew in the clock signal paths (the difference in delay between circuits receiving the same clock signal) becomes significant. A layout system should, therefore, automatically insert clock buffers to maintain even skew across the clock tree network.

A related issue is the proper sizing of cell output buffers in accordance with their load-driving requirements. Typically, a cell library will contain several versions of the same cell, but with different output drives. The user must make educated guesses as to which is the most appropriate for performance and area optimization. More sophisticated tools provide an automatic buffer-sizing capability, freeing the designer from such concerns.

The ASIC Design Environment

A complete design environment encompasses all the tools necessary to complete the design of an ASIC. The necessary resources include cell libraries, tools for design capture, simulation, timing analysis, layout, packaging, test, and documentation.

Factors influencing the selection of an ASIC design environment include the following:

- Type and complexity of ASICs to be designed.

- Design methodology.

- Number of ASICs to be designed/project/year.

- Expertise of the engineers using the tools.

- Level of ASIC vendor/user interface.

- EDA tools budget.

In any design environment, a minimum set of tools will be required to do the job at the anticipated level of vendor interface. This minimum set may accomplish the design task, but may do so without regard to the quality of the ride. Other tools, depending on the application, may be considered required or optional equipment. Accelerators, emulators, and certain layout and verification tools may greatly enhance productivity, but their costs may be prohibitive.

Since most design departments are not blessed with an unlimited budget, care must be taken to satisfy design needs within the allocated budget. The acquisition of tools and equipment may involve a combination of financing arrangements, including purchase, lease, rental, and time share. The total design system cost, though, should be weighed against productivity gains and overall return on investment.

Alternatives to an in-house design environment include the ASIC vendors' field offices, which are often equipped with design center facilities. There are also independent, third-party design service centers that have a variety of tools and may represent several ASIC vendors. Distributors may also provide design center facilities and services for the ASIC vendors they represent.

For many ASIC designers, a schematic capture system is the only CAE tool available in-house. Used in conjunction with the ASIC vendor's library, it produces the netlist, which is later used for simulation and layout. With the exception of vendor-performed turnkey designs, unsimulated netlist submission represents the lowest level of user/vendor interface. Such users rely on the ASIC vendors' resources for performing simulation on a computer time-share basis, or they may have the vendor perform simulation. The costs associated with such services, though, can often become prohibitive. In many cases, the computer charges exceed the prototype NRE. Another drawback is that engineers must learn a new set of design tools for every ASIC vendor they may use. Costs, lack of design security, travel, and location inconveniences are other concerns. For these reasons, the user should seek to bring as much EDA capability as possible in-house. The return on investment after just one project may be justification enough.

Maintenance and Upgrades

Maintenance, which covers bug fixes, revisions of the current release, and technical support, typically costs 1% of the list price per month. It is generally paid on a yearly basis, but it may also be paid quarterly or monthly.

The costs of upgrades to the next release is another cost item that must be accounted for. Some EDA vendors will support one release back, but many will require the upgrade fee in order to continue maintenance services. In addition to new features, capabilities, and bug fixes, new software releases inevitably have bugs of their own. Also, the upgrade may impose the need for additional memory or an operating system upgrade as well. In fact, some users, while in the middle of a design project, will refuse the upgrade to avoid the disruption and the potential risks associated with new bugs. The bugs these users may have encountered in the course of their design work would typically have been addressed through standard maintenance services such as the provision of work-arounds and patch fixes.

Design Tool Integration

Identifying the best point solutions (the best available tool for a given application) is one thing; making them all play together is quite another. Since no one company can provide all the tools the user needs, CAE support personnel must often spend valuable time making different tools work together, mixing and matching in-house and commercial tools and integrating them in a manner that is transparent to the users.

Typically, CAE administration personnel will purchase *framework* software, plus a core set of tools from a major CAE vendor and use in-house or third-party tools to complete the design environment. Gluing together such diverse tools, though, presents some serious difficulties. These have included the lack of a consistent user interface, the need to learn and manage multiple tools, the lack of translators or interface utilities, and the inability to deal with multilevel design models. A number of solutions are emerging, though, effectively moving much of the burden from the users to the CAE vendors, who are beginning to assume the responsibility of making their tools more open and consistent. Standards for CAE tool interfaces and data formats promise to help improve the situation, but they will not solve all the problems in the near future.

EDIF

One solution to the design data translation problem is the standard Electronic Design Interchange Format (EDIF), which is a universal, neutral format for transporting data created by one design tool to another. Obviously, the use of EDIF requires that each tool in the design environment have a translator to the format.

A design tool that claims EDIF compatibility may mean it is able to write EDIF files, read EDIF files, or write and read EDIF files. Additionally, the EDIF interface may support only a limited number of circuit views. The exact capabilities of the tools' interfaces, therefore, should be well understood.

Frameworks

Design data management has become increasingly important as teams of engineers are involved in the design of more complex systems. Maintaining different representations of the same data in order to accommodate differences in tools can become a nightmare.

The concept of a design tools framework (Figure 5.6) involves a common database format that allows the transfer of design data from one tool to the next without error-prone translation. It is based on a single, object-oriented database in

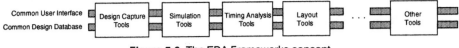

Figure 5.6 The EDA Frameworks concept

which a single data model integrates each view of the design data: schematic, symbol, simulation, and layout. The design database is independent of the tools used to generate it. As long as the tools can read and write the common database format, they can easily be integrated into the design environment. Interface at the database level allows more information to flow from one tool to another than is possible by passing only a netlist. Another frameworks feature is a consistent user interface, which provides a familiar look and feel for each tool. Generally, the interface to all the tools is graphical, or icon-based, so an engineer can use the different tools without having to remember individual tool invocation sequences and operating commands.

CFI (CAD Framework Initiative) is a consortium of companies whose charter is to define framework guidelines and standards. Until the standard is adopted industry-wide, though, a number of companies will continue to put forth their own concept of what the standard should be. Currently, many CAE vendors are integrating their design tools with proprietary and third-party framework offerings, as well as providing frameworks that their own and outside tools can plug into.

For users with existing design environments, migrating to a frameworks-based environment can pose some difficulties. Because of the considerable investment they may have made in existing design tools, users cannot simply scrap everything and start over. Rather, the migration will likely have to occur in stages. Also, it may be some time before all major EDA vendors will support a standard framework environment.

Workstation Hardware

Although it seems there is a veritable plethora of platforms on the market, almost all CAE vendors run under UNIX (be advised of the different versions) or DOS on such platforms as Hewlett-Packard (formerly Apollo), Sun, Digital Equipment Corp., or IBM.

The price/performance developments of these workstations has been staggering. Only ten years after their initial introduction, workstation prices fell from $50,000/MIP to $500/MIP. This trend is continuing, with performance doubling at half the price every 12 to 18 months. At this rate, the selected workstation will rapidly become obsolete. Therefore, it should be purchased with the expectation that it will likely need to be replaced in two to three years.

The base price of an engineering workstation can be very misleading. When evaluating the cost, it is important to look at the total system cost, including disk and cartridge tape drives, memory, network resources, other peripherals, maintenance, and system administration. Managing the tools, the project, and the data can add more than 30% overhead to the design process.

The workstation configuration should be able to support the types of designs to be done. For example, a 10,000-gate ASIC may require as much as 150 MB of disk storage and a minimum of 8 MB of RAM to perform even simple simulations. The disk and RAM requirements scale almost linearly with increased chip complexity. Also, as chip designs become larger and more complex, the workstation performance of simulation and other tasks degrades significantly. For this reason, hardware accelerators should be a serious consideration.

Design Tool Quality

For products like semiconductors, measuring quality is straightforward. Either the product meets spec and passes or it fails. The overall quality level can be measured in terms of defective parts per million (ppm). Software, on the other hand, does not lend itself to such empirical metrics. Rather, quality is measured in more subjective terms. A product's quality is said to be good if it performs to the customer's expectations. This may sound simple, but when one considers the infinite combinations of hardware configurations, software, applications, and user expertise, measuring quality in absolute, objective terms is nearly impossible. The measures of software quality necessarily shift to more subjective evaluations of such things as timely bug fixes, frequency of system crashes, application support, and customer service. Poor software quality can add days and even months to the design schedule.

Unfortunately, most quality assurance activities are performed by the users. Due to the complex nature of ASIC design tools, as well as the EDA vendors' time-to-market pressures, new product releases will inevitably contain bugs. Some are the result of poor quality control; some are knowingly shipped.

Typically, a new release goes through several phases before it is deemed *production ready*. After the vendor's testing is completed, it is usually released to a handful of special customers who are eager to get their hands on the latest tools, even though they may contain bugs. In return for getting the release early, they agree to run the product through its paces in an actual design environment, record all bugs, and make suggestions for improvements. The recommendations of these "beta site" customers may then be incorporated into the product, tested, and finally released as production software. As more customers use the product, more bugs will be uncovered. Work-arounds or fixes are generally supplied on an individual basis. All reported bugs are accumulated by the vendor until it is appropriate to issue a maintenance release, a revision of the software that incorporates all the bug fixes to date. And so it continues until it becomes necessary to issue a new release, which may include a substantial number of new features, thus starting the cycle over.

All too often, however, a customer becomes an unwitting beta site. Promises of delivering certain unreleased capabilities may be committed by the vendor in order to make a sale. Therefore, it is important for the user to understand the status of the software that he will be receiving. It is also recommended that one talk with several reference accounts to obtain more objective opinions. The following list of issues provides a good start:

- Length of time the system has been in use.

- System configuration (host workstation).

- Type of designs (technology, methodology, complexity).

- Number of designs completed.

- Number and nature of software problems.

- Average response time to provide a fix or work-around.

- Strengths and weaknesses of the system.

- Level of overall satisfaction (tools, ease of use, customer service, application support).

- Time required to learn and become productive.

In just a few short years, the EDA industry has grown from essentially three suppliers (Daisy, Mentor, and Valid) providing turnkey design systems to more than a hundred smaller companies providing niche-oriented tools focusing on specific design tasks. With so many diverse offerings, the selection of the EDA system can be extremely complicated. But many good tools are available. The key to finding the right system is attention to detail, clear understanding of current and future needs, and a lot of hard, tenacious work. Although selection is a process of elimination, the experience can be made easier by focusing on established market leaders whose directions are clear and who offer long-term migration strategies for their users. Support of de facto or industry standards that enable a user to tie diverse systems and applications into a single environment is also a critical consideration.

The point at which most EDA tools shoppers begin is the demonstration. Be advised, though, that most product demonstrations are prepared and polished, and the system can be shown to do things that are not as easy to perform in an actual design environment. The demonstration is only a starting point. The next level of evaluation should be based on specific benchmarks that will be typical of the user's own applications. Depending on the nature of the benchmark, the vendor may be willing to perform the work. Vendors will also often install an evaluation copy or provide the system under a 30 to 90-day unconditional guarantee or under a conditional purchase order.

Finally, a detailed quotation outlining total system costs, installation costs, training, maintenance, and upgrade costs, and recommended hardware configuration will need to be obtained. Costs are generally provided on a per seat basis. That is, software is licensed for use on a specific workstation or node. Movement of the software from one node to another may, therefore, be restricted. If the design environment requires multiple seats, a site license may be a more cost-effective approach.

Chapter 6

Selecting the ASIC Vendor

THE SELECTION OF AN ASIC VENDOR can be a complex and time consuming undertaking. It's a tedious process of investigating the services and capabilities offered by numerous ASIC vendors and determining how they can best accommodate the specific needs at hand. Vendor selection is best approached through a process of elimination. The specific requirements for the ASIC are a first-order filter that can narrow the field of contenders. The selected vendor's approach must be consistent and compatible with the desired technology, methodology, and design tools. The vendor must also be able to accommodate the desired level of design interface.

Vendor selection is perhaps the most important activity in the ASIC design project. The ASIC vendor will, in effect, be a codesigner of the ASIC, and the combined performance of both parties will determine the success of the project. A great many vendors compete for a relatively few number of new designs. This is an advantage that should be leveraged in vendor selection and negotiation. The wrong vendor can severely jeopardize the entire project.

Preparing the Request for Proposal

Vendor selection should be based on a combination of objective and subjective criteria. The communication of these criteria is best performed through the use of a formal request for proposal (RFP).

The RFP is the procurement specification for the ASIC. It is used to document the proposed ASIC, as well as to qualify potential suppliers. It is the mechanism by which ASIC vendors evaluate the ASIC requirements and prepare a specific proposal for its design, accounting for all the associated costs. The more complete the RFP, the better it will communicate the requirements, thus minimizing any potential misunderstandings. Furthermore, the more detail provided, the more accurate the vendors' proposals will be. To eliminate surprises in the middle of the design cycle, all expectations must be well understood by both parties, up front. The RFP should, therefore, be a formalized document, reviewed by all members of the system design team to ensure that all the requirements and specifications have been comprehended and interpreted correctly.

The RFP is generally divided into two sections. The first describes the pertinent details of the design. How much information to provide is a discretionary matter. The information exchange may be informal or under terms of a nondisclosure agreement (NDA). Any confidential information should be identified as such by stamping all affected documents "CONFIDENTIAL AND PROPRIETARY." The terms of distribution should be on a need to know basis, with no copies to be made. (Submitting sensitive information on red paper makes it difficult to photocopy.) The return of all materials provided for quotation purposes should also be requested.

The second part of the RFP directs the vendors' response. It solicits a specific proposed approach to the design, and provides any special instructions to the bidder. There are several classes of RFPs; the choice of which to use depends on the nature of the development program and the status of the ASIC specification. They are as follows:

• Request for information (RFI). The RFI is used to solicit data for a feasibility study. In this case, there is more interest in whether a development is possible, rather than what its cost will be.

• Rough order of magnitude (ROM): The ROM is used when ball-park pricing is needed for a design that is subject to change or has aspects that are yet to be determined.

• Budgetary: The budgetary quotation is the most common type of RFP. It is used when the ASIC is fairly well defined. Quoted prices are fairly accurate estimates and are generally tied to a set of assumptions, which may be subject to change as the development proceeds. The vendor is not contractually bound by budgetary pricing.

• Not to exceed (NTE): An NTE quotation is used when budgets impose a cost ceiling for the ASIC project. Lower prices may be negotiated, but the vendor is contractually bound only by the NTE pricing.

• Firm fixed price (FFP): An FFP proposal can only be expected when the ASIC specification is finalized and frozen. The pricing is contractually guaranteed by the vendor.

The following provides guidelines for preparing an RFP.

Contact information: This introductory section identifies the company, the address, phone, and fax numbers, as well as the appropriate technical and administrative contacts.

Design tool requirements: If a design tool environment has been established, identify those tools and data formats that the vendor must support.

Design interface level: This section describes the basic approach to the design task. It includes a statement of work (SOW) and assigns the responsibilities to be performed by each party. It should also define the data transfer points and a description of the data (schematic, netlist, physical design data) to be provided. If the design is to be performed *turn-key* by the vendor, describe the design documentation to be provided.

Functional description: A brief functional description of the circuit and how it will interact with the system is important in order for the vendor to interpret the design's block diagram and schematics.

Functional requirements: List all macro functions to be included in the ASIC. These would include RAM, ROM (their organization and access time requirements), register files, multipliers, peripheral macros, and analog functions. In addition to these functions, include an estimate of the number of gates to be used for combinatorial logic. List also any special design constraints, such as requirements for scan latches, clocking schemes, etc.

Performance requirements: State the system clock frequency, and if and how the clock is divided down on chip. Critical paths should be clearly marked on the schematic. Describe what must happen within a single clock cycle and what, if anything, can be pipelined. Identify whether the path is internal to the ASIC or whether it must include I/O delays as well. Define the conditions under which critical timing must be met (voltage, temperature, process corners, output loading, and others).

I/O requirements: A pin description list is the best method for conveying this information. It should include the following:

- Total number of pads required

- Total number of inputs

- Total number of outputs

- Total number of bidirectional pins

- Total number of power and ground pairs

- Output drive, load and logic interface level for each pin

- Total number of outputs simultaneously switching and/or driving

- Any special requirements, including ESD protection levels

Package requirements: Depending on the application, ASICs can be provided in one of three forms: wafer, sorted die, or packaged. If packaged, state whether the package is to be surface mounted or through-hole mounted, plastic or ceramic. If a specific package has been identified, include a drawing. Describe the environment and thermal conditions (chip power, airflow) under which the ASIC will operate. List any pin-out requirements, power issues or concerns. Specify, as appropriate, the use of shipping carriers, trays, or tubes. If there are different requirements for the prototype and production packages, specify that as well. It is often preferred to receive the prototype units in ceramic packaging with a removable lid in order to facilitate debugging (allows inspection and detection of wirebonding or pin-out errors).

Methodology and technology requirements: Design objectives (performance, level of integration), for the most part, dictate the selection of design methodology (custom, cell-based, array-based) and technology (CMOS, BiCMOS, ECL, GaAs, rad hard). The process feature sizes required to satisfy these objectives may be less obvious. One vendor may require the use of a leading-edge process to satisfy the design objectives, while another may be able to use a less aggressive, but more mature process, by virtue of its superior design tools. If a relatively new, leading-edge process is required to satisfy the performance requirements as specified, it may be wise to reconsider the system partitioning to take advantage of a lower-cost, lower-risk mainstream technology. If critical path timing is a problem, consider pipelining the design (by adding intermediate registers and clock cycles inside the path) to accommodate it in a less aggressive process.

Absolute maximum ratings & operating conditions: Specify the following:

- Operating temperature, storage temperature

- Supply voltage range

- Maximum power dissipation

Production requirements: State the number of prototype units needed, the preproduction quantities, and the anticipated range of annual production unit quantities. Estimate also the year and quarter that production is to begin and what the life of the product is expected to be.

Testing and environmental screening requirements: Estimate the number of test vectors required to test the ASIC to the desired level of fault coverage, as well as all dc parametric, functional, and ac test requirements. State also the test limits and conditions. Define the level of test that determines prototype acceptance. Indicate whether serial scan testing is to be performed. If there are any special tester requirements (number of pins, tester accuracy, resolution, and clock rate), list these as well. If any environmental or MIL-STD screening is required, list the applicable specifications, test groups, and methods, as well as any applicable modifications or waivers to those specifications. If burn-in is to be performed, specify the temperature, duration, and bias conditions (static or dynamic).

Design documentation: Provide all necessary block diagrams, schematics, waveforms, timing diagrams, applicable documents, specifications, drawings, and standards as attachments.

Deliverable items: Identify all those items whose delivery constitutes fulfillment of all contractual obligations. These typically include a minimum number of tested prototype units, design documentation, netlists, plots, datalogs, test programs, design verification results, and final reports.

Quality assurance provisions: Outline all quality assurance requirements, including responsibility for inspections, lot sizes, lot acceptance criteria, quality conformance testing, nonconformant material disposition, and any applicable quality conformance specifications and standards.

Instructions to the bidders: Include any special instructions as to how the response to the RFP is to be structured. Specify that costs are to be itemized and all-inclusive, comprehending all special engineering and tooling that must be designed, manufactured, or acquired. The proposal should also include a schedule and milestone chart, any assumptions upon which the proposal is based, any potential problem areas, and suggested alternative approaches. If a vendor qualification questionnaire is included, provide instructions for its completion as well. Include also a deadline for submittal and a proposal validity period.

Evaluating the Bidders

Vendor evaluation encompasses the response to RFP and questionnaire, the presentations, demonstrations, benchmarks, and the quality of the local support. Keep in mind also that there is an inevitable tendency for vendors to inflate the capabilities of their products, combined with their tendency to downplay costs.

Unfortunately, there are no magic formulas available for selecting the best ASIC vendor. Each project is unique and some vendors will be better equipped to handle a particular type of design. The specific profile of a supplier will determine how appropriate it might be for a given application. However, there are no panaceas either. ASIC design is a process of trade-offs, and that includes deciding which vendor to use. The key is to understand the vendors' strengths as well as their deficiencies and how the inevitable shortcomings will be compensated for.

The task of sorting everything out can be overwhelming. To manage all the data and maintain a level of objectivity, a vendor evaluation matrix is in order. A matrix can be a tremendous aid in the selection of the best qualified vendor according to their performance to the various evaluation criteria. Each vendor's score can be determined by assigning a weight or importance factor to the rating in each of the evaluation parameters and calculating the total score. The results provide an objective, quantitative data point to use in conjunction with more subjective evaluation criteria. The supplementary programs described in Appendix A include a PC-based evaluation matrix that automatically determines the best vendor, based upon the user's input.

When reviewing the RFP responses, look for completeness, sound technical solutions that reflect an understanding of requirements, and pricing that is reasonably justified. In addition, the following review of other issues and concerns can be used to help complete the evaluation, as well as to identify and minimize potential risk areas.

Technical Issues

Is the vendor able to satisfy all design specifications? How comprehensive is the library? Have all the cells required by the design been previously fabricated and characterized in the target process? Will any custom module development be required? Some custom development merely requires netlist modification involving existing, precharacterized cells. Others require a completely new, fully custom layout, which could impose a great deal of risk.

Does the vendor offer a variety of processes and methodologies? How mature is the applicable process? Has the vendor completed a similar chip of similar complexity in a similar package? Can the design be easily migrated from one design methodology to the next (gate array to cell-based), as well as to the next generation process? What is the level of effort required and what are the associated costs?

Are all critical timing requirements satisfied? All accompanying benchmarks should include not only the intrinsic delay of each module in the path, but also the delay due to the parasitic routing capacitances. The accuracy of the gate-level simulation models should be compared with a SPICE circuit simulation of the same path. This will ensure the accuracy of the delays, as well as qualify how well the gate-level models have been calibrated.

Will the vendor guarantee the accuracy of its workstation models? Will the vendor guarantee chip performance at the system clock rate? (Many vendors will only guarantee dc parameters and functionality at a low-speed tester clock rate, such as 1 MHz).

Cost Issues

Obtain a detailed cost breakdown of all items included in the NRE. Identify the fixed and variable cost line items. For variable costs, negotiate a not-to-exceed (NTE) cost limit. Determine the costs for iterations on any of the line items that may require iteration. Quite often it is necessary to tweak a design after the first silicon samples have been evaluated. What are the costs per mask layer changed if revised prototypes are needed?

Production unit pricing is based on the required process, the die size, package, and volume. The die size is the variable at greatest risk, since it must be estimated. If the chip size increases to the point that it will no longer fit into the target package cavity, has an alternative package been identified or will the package need to be retooled? Will the vendor bear this risk? Also, the production unit cost can increase sharply with small incremental increases in die area. Will the vendor guarantee its die size estimate and therefore the unit pricing?

Design and Prototyping

Can the vendor provide all the necessary design tools, design services, manufacturing, packaging, and testing? Are subcontractors used to perform any of the activities? If so, what is the nature of their relationship? How does the vendor monitor or guarantee the quality levels of their subcontractors' processes? Will the vendor provide technical and procurement contacts for each of its subcontractors to be used in the execution of the ASIC design?

What is the vendor's ability to meet prototype delivery schedule? Does the vendor provide an option for accelerated processing throughput? Such services typically carry a premium of $20,000 or more but may shave weeks off the prototype development schedule. Is the schedule realistic? Has it comprehended the availability of needed design tools, development of custom cells or modules, package and socket tooling and lead time, test development and debug? Simulation is the task most commonly underestimated, in terms of the time needed to complete it. Can the vendor provide assistance in assessing the time and resources needed to complete each design task?

Who will be responsible for physical design (layout) and back-annotation of the simulation files? If these activities are to be performed by the vendor, what is the cost of each iteration not comprehended in the quoted NRE?

What level of design support is included in the NRE? What experience does the vendor have with similar designs and systems? How well do they understand the application? Can they add any value in the architecting and partitioning of the design?

Design Tools

Are the tools easy to use and well documented? How mature are they? What are their costs? How flexible are the terms? Can the tools be leased or rented? Depending on how attractive some business may be, the vendor may provide all the necessary design tools at no cost or at the price of maintenance only.

If the vendor's proprietary design tools are to be acquired, will they be compatible only with that vendor's processes? If the vendor uses third-party foundry services, does the foundry use the same set of tools? Have they qualified them for use with their processes?

What third-party tools are supported? Will the desired design environment be supported? If translators or interfaces must be developed, how will they affect the schedule? Any tools or interfaces developed in support of the project will, by their very nature, be unproved, thus imposing substantial risk on the project.

Production Issues

If the vendor does not have its own wafer fabrication facility, but uses third-party foundries, what is the nature of their relationship? Is there a foundry agreement in place? Will the vendor provide a copy? Can a written statement from the foundry be obtained that commits their support for the ASIC vendor as well as the ASIC production supply? How were the foundry and its processes qualified? Is characterization data available on test and qualification devices? How is the vendor/foundry interface performed? How long has the relationship been in effect? How many unique chip designs has the vendor processed there? How are the supply, quality, and pricing guaranteed? Does the foundry perform a true foundry service or is it simply selling its excess capacity? How will pricing and delivery be affected when demand for the foundry's standard products ramp up? Will they continue to support small production volumes of a large variety of unique chips? Wafer fab facilities are routinely bought and sold. If the ASIC vendor's wafer source is sold, how will supply be affected?

What was the vendor's on-time delivery performance during the past two quarters? How many unique ASICs did it include and in what volumes were they shipped? Is there a minimum-lot quantity requirement? Is it based on line item units or dollars per line? What is the cost for additional prototype or preproduction units? What is the production lead time? Can the vendor facilitate "hot lots"? If the ASIC is to be prototyped on shared, multiproject wafers, will a new dedicated mask set need to be made to support production requirements? Have those costs been amortized into the production unit pricing? Does the vendor offer yield enhancement and product engineering support services? Will any resulting yield improvements be reflected in lower prices? Will the vendor guarantee long-term price reductions so that the product can continue to be competitive? Does the vendor have an active second source?

Business Issues

There are a great many ASIC suppliers from which to choose. For the most part, however, ASICs have been a major revenue source for a relative handful of suppliers. Of the 100 or more ASIC suppliers in the world, only about 15 or so have managed to grow the ASIC segment of their business past the $100M mark. Many of these have operated with short-term losses as they vacillate between positions of increasing profit margins or building market share.

Are ASICs the primary thrust of the company? The strength of broad-based companies is that they offer a wide variety of process technologies, fabs, design tools, and other resources. They also have large staffs that typically include experts in each of the many disciplines. The extent to which these services and capabilities are required depends, of course, on the specific needs, internal resources, and experience of the user. There are also downsides to working with large, broad-based companies. They simply may not be as responsive and supportive. If the production volume requirements are low, they may not be interested in the business. The smaller, niche-oriented suppliers, though, may offer superior technical solutions and better service and may also be able to provide the same quality and delivery at a lower cost.

Is the vendor profitable? Consult the vendor's annual report for financial information, levels of investment in R&D, manufacturing and ASIC emphasis. The supplier should complete at least 100 different ASIC designs each year. There is a significant learning curve in ASIC design and the higher the rate of designs, the higher the likelihood that problems in procedures, processes, tools and libraries have been ironed out.

What is the vendor's market position relative to its competitors? If a candidate vendor is not among the top ten, it should have a clear technological superiority. If a second- or third-tier supplier has nothing unique to offer, it could be an unnecessary risk and should be avoided. If a smaller company is selected, assess its long-term viability. Does it have the financial resources to maintain and improve its competitive edge?

Obtain a minimum of five references—both good and bad. Although vendors will be reluctant to provide references of less than perfect projects, it is equally important to qualify how the vendor deals with problem situations. Did they work diligently with the customer to solve the problem? What was the nature of the solution?

Contracting with ASIC Suppliers

The contract is the controlling document constituting the complete understanding between the parties, and it defines the terms and conditions under which the business will be conducted. The contract is as binding as its provisions make it to be. Generally, the contract is based on the best efforts of the parties in their respective performance. The contract sets forth the responsibilities and deliverables by each party and defines what constitutes fulfillment of obligations. It may include any desired terms, such as long-term fixed price arrangements, royalty rights to proprietary circuitry, or any other arrangement. Typically, both parties will have their own set of terms and conditions. Inevitably, both parties will have exceptions to certain terms, and compromises must be worked out. Most contracts, though, will include language describing or governing each of the following areas:

- Statement of work and milestones

- Price and payment terms

- Delivery

- Proprietary rights

- Ownership

- Reference documents, specifications, schematics, drawings

- Nonconformance corrective actions

- Termination

- Waivers, disclaimers, and liability limitations

- Patent infringement

- Warranty

A few areas warranting additional detail are proprietary rights, ownership, and termination. Usually, the customer must acknowledge that the photomasks (and the tapes used to generate them) and finished goods contain information proprietary to the seller. These include the design and layout of the cells used in the design, among other things. These items may involve patents, copyrights, and trade secrets. As such, the vendor will claim ownership of all tapes, masks, and programs utilized in the execution of the contract. The vendor will acknowledge the proprietary rights in the customer's end product design, that is, the manner in which the cells have been interconnected to perform a desired function. Under such terms, the customer is not free to take the mask set to another foundry and have the parts fabricated. The contract will generally stipulate, though, that in the event that the supplier defaults,

109

the ownership transfers to the customer. To make the terms more effective, a copy of the mask generation tape can be placed in escrow. If for some reason the masks cannot be made available to the customer, a new mask set will have to be generated and the parts can be fabricated at the appropriate second source. Because a complete mask set can cost between $20,000 and $30,000, it is not practical to hold a duplicate set in escrow. (Second sourcing at the mask level is the exception, and not the rule in the ASIC industry. If this is a requirement, it will limit the number of candidate ASIC vendors.). In the event of nonperformance by either party, for whatever reason, the contract can be terminated with proper notice. Financial liability for work performed prior to the notice of termination, however, will generally remain.

Chapter 7

Design Guidelines and Issues

THE THINGS THAT CAN GO WRONG in an ASIC design are numerous, but if one pays careful attention to proper design practices, testability considerations, and common problem areas, the chances of achieving first-pass success are much improved.

The first steps in the ASIC design process include determining the most efficient and cost-effective circuit partitioning and developing a comprehensive specification from which all design and analysis work will follow. Effective project management will steer the design through its many phases, keeping the project on track, on schedule, and within budget. With these objectives in mind, the following material provides an overview of the issues that can make or break an ASIC design project.

Partitioning

Partitioning is the process of dividing a system into sections that can be implemented at the level of the target technology. The system is partitioned into boards, each board is partitioned into logical sections, and each section is partitioned into individual components, some of which will be ASICs. The process can be as simple as drawing boundaries around the logic that would appear to fit into the ASIC.

Prior to the start of the partitioning task, the system objectives and available implementation technologies must be well understood. Partitioning is a trial and error process based on the desired functionality and limited by process, die size, package, and I/O constraints. Partitioning will inevitably involve a series of trade-offs with regard to what functions should and should not be included on the ASIC. It must be decided whether to include memory, core microprocessors, and other macro cells in the ASIC or leave them off-chip in their standard, off-shelf form. Sections of the circuitry that are likely to change or require updating should be isolated and left off-chip altogether.

When a given circuit partitioning result becomes too large for the available package or array or is no longer feasible for economic reasons, either certain functions can be put off-chip or the circuit can be partitioned into multiple ASICs. In a multichip partition, it may be cost effective to repeat logic in each chip in order to reduce I/O pins. Also, as I/O counts increase, the chip can become pad limited. Such designs result in a considerable amount of wasted silicon, which can increase the production unit costs. It may be desirable, then, to integrate other board functions into the ASIC, space permitting.

If the ASIC will serve the needs of more than one end product, functions specific to only one of the products can be included and enabled via control or mode pins or by wirebonding the appropriate pads for each design option. This will result in one ASIC that may serve two different functions, instead of requiring two unique ASICs. The higher production volumes for the one device type will also result in lower unit costs.

ASIC partitioning is most intuitive when it is defined along functional lines. If possible, it is helpful to think of the design in terms of functional modules, or blocks. A review of the system block diagram is a good starting point for this approach. Testability concerns also have a hand in determining an optimal partitioning. The architecture, no matter what the final partition looks like, must allow adequate test access.

Partitioning must also respect circuit timing considerations. Drawing the line between data path elements, for example, can result in timing violations due to the on- and off-chip delays and clock skew. High-speed designs, therefore, require careful analysis of chip-to-chip delays to ensure an efficient partitioning scheme.

Dealing with wide buses can also present complications for partitioning, due to the large number of I/Os potentially running between chips. One possible solution is to segment the circuit into a bit slice architecture, where bus bit widths/chip can be reduced, thus simplifying I/O concerns. Multiplexing I/Os can also alleviate I/O partitioning problems.

Including Core Functions on the ASIC

The use of core functions, such as microprocessors, microcontrollers, memory, and peripherals, can facilitate the integration of entire systems on a single chip. The higher levels of design integration made possible by submicron processes make it possible to mix such a diverse assortment of circuit elements.

In the case of core microprocessors, they contain all the logic elements needed to execute its instruction set. These generally include microcode ROM, state machines, a program counter, registers, an ALU, and other required logic. Such cores are commonly used as enhanced microcontrollers, based on such industry standards as the 8051. The use of these cores is facilitated by the availability of inexpensive development systems and emulators for developing and debugging system software. Another advantage of core microprocessors is that memory and peripheral functions can be connected directly to the address and data outputs of the processor, thus optimizing system performance.

In some applications, it may be necessary to modify the functionality of a standard core microprocessor. The degree to which it can be modified, however, depends on how the core was originally designed. A core, or macro, is said to be *hard* when it is based on an exact layout copy of the standard component. Because it is defined at the layout level, it cannot easily be modified or migrated to newer, more advanced processes. Unless the standard part is redesigned in a newer process, it remains frozen in the original design rules. As a result, the size of the cores (which must be fabricated in the original or compatible process) may severely limit their use in an ASIC, due to the unacceptably large resulting die sizes. The advantage hard macro cores have, though, is that they are very well characterized and their performance parameters are well documented.

Soft macros, on the other hand, are defined at the cell library and netlist level. As such, they can be easily modified and migrated to new processes without requiring a complete redesign. Of course, if a core is modified, the vendor may no longer guarantee its functionality.

In spite of all the advantages offered by embedded cores, their inclusion in an ASIC can present some serious complications for the design. Problems arise in the areas of simulation, test, and economic justification.

The simulation of microprocessors and other complex cores is often complicated by the unavailability or the unwillingness of suppliers to provide accurate gate-level models. The use of a core microprocessor in an ASIC will also substantially increase

113

the complexity of the simulation task. To fully analyze a core-based design, the simulator must be able to work at the instruction-set level as well as the behavior, register-transfer and gate-levels. Simulating at the instruction-set level is critical because it allows the system software to be run on the simulator, which is an extremely useful debugging aid. Also, to facilitate the debugging process, an emulation capability will be required that includes the ability to display and change register and memory contents, as well as to set breakpoints and single-step instructions.

Testing an ASIC with an embedded core function can also be difficult. Because vendor-supplied macro cores are proprietary designs, detailed gate-level models may not be available. This can make it difficult, if not impossible, to attain high fault coverage. To compensate, most vendors will supply test patterns for their cores. The extent of fault coverage these patterns provide, though, should be understood. Some suppliers will also provide the special circuitry that may be required to isolate and test the core.

Although the availability of core functions is increasing, they are still generally far more costly when included on the ASIC than when purchased as standard parts. An off-shelf 8-bit microcontroller, for example, may cost $2 in volume, but the same function on an ASIC may boost the chip cost by $10 to $20 or more.

Mixed-signal Design

As ASICs replace more and more digital circuitry, discrete analog components become an increasingly significant contributor to overall system size and cost. The integration of these analog functions, therefore, is becoming a priority for many applications. Most electronic systems have some analog content that are potential targets for integration. In fact, some industry analysts anticipate that more than half of all ASIC design starts will include some type of analog circuitry. The advantages of integrating analog functions on-chip are similar to those for digital functions. The resulting ASICs trim costs, occupy less board space, and consume less power. But even without analog functions on chip, as process geometries continue to shrink, parasitic effects on high-speed logic are becoming more significant, and the design of these chips is taking on more of the attributes of analog design.

As desirous as it may be to gather analog functions into ASICs, there are major obstacles that would-be designers of mixed-signal ASICs will inevitably face. These include limited available processes, a general lack of adequate design tools, and very difficult test methodologies. Selecting the most appropriate supplier for a mixed-signal design is also very difficult because, while one supplier may have the best process, another may have the best tools. Another important issue is the decision of whether to design the chip in-house or have the vendor perform the design, especially if in-house analog design expertise is limited.

Because of the difficulties in simulation and potential innaccuracies of analog models, the design of mixed-signal ASICs is an inherently more costly and risky undertaking. Most require multiple fabrication iterations. The first pass is typically used to gather characterization data so that the design can be fine tuned, as it is particularly difficult to model analog performance accurately over temperature, voltage and process variations. Maintaining manufacturing repeatability is another major concern.

The lack of adequate design tools has been the major stumbling block to more widespread integration of analog and digital functions. In addition to a multilevel, mixed-signal simulator, mixed-signal designs also require specialized layout tools that are peculiar to analog design constraints. The primary provision is that analog circuits must be isolated from digital switching noise. In a purely digital circuit, the primary concern of physical design is controlling the effects of capacitance and resistance on circuit delay. When analog functions are included, the problems expand to include controlling crosstalk between adjacent nets. Close proximity of noisy digital signals to sensitive analog circuits can render the ASIC inoperative. To avoid these problems, analog circuits and signal lines must be separated from those of the digital section. In many cases separate power supplies must be provided for the analog circuitry to further isolate them.

Mixed-signal Simulation and Test Issues

SPICE (Simulation Program with Integrated Circuit Emphasis), the most commonly used circuit simulator, works with precise diode and transistor models to characterize devices and perform circuit analysis. The program is in the public domain, and many EDA vendors offer a version of it that is integrated with their own tools or is easier to use. SPICE, which is extremely compute intensive, is the primary tool used for the simulation of analog designs. However, because of the time required to run an entire chip simulation at the transistor level, it is simply not practical for most applications. If the digital portion of the circuit is small enough, though, SPICE can be used for the entire circuit. Otherwise, the best solutions offer an ability to perform simulations that are able to mix transistor, analog, and digital behavior and gate-level models. The majority of mixed-signal simulators *glue* the different simulator types together. The digital portion of the design is simulated with a standard logic simulator, while the analog portions of the design are simulated with SPICE. The interface between the two simulations represents the real challenge—particularly if the design contains feedback between the digital and analog circuitry. Newer simulators, however, are overcoming many of the problems and no longer require that the design be split into digital and analog sections for analysis. Although these systems are typically glued, they allow the effective exchange of intermediate results between the two simulators. The key to using the tools successfully, though, depends on the availability of accurately modeled mixed-signal ASIC libraries. And models represent another area of trade-offs. Behavioral models, whether for analog or digital cells, execute very fast. SPICE, or transistor-level models, on the other hand, are extremely slow in executing, but are very accurate. As a compromise, many suppliers offer *digital-equivalent* models for their analog cells. These models can easily be used in a standard logic simulation where verification of gross analog functionality is all that is required.

Mixed-signal ASICs can run into problems in the test area as well. For purely digital designs, the test program is developed, somewhat automatically, from the simulation patterns. There is no easy way, however, to automate the analog test conversion. If any precision analog measurements are required, two test programs for two different testers (one digital, the other analog) may be required to test the device. Analog test development and testing is very labor intensive, slow, and expensive compared to digital testing. Analog testers are also limited in the pin counts they can support. Furthermore, having to test a device on two different testers is an extremely undesirable circumstance. The test capabilities of the ASIC vendor, therefore, should be a major selection criteria.

The tester must support analog waveform generation and have precision voltage and current sources, voltmeters, and ammeters and, as a rule of thumb, should have a resolution of at least one magnitude better than the signal being measured. Digital testers are generally inadequate for testing analog circuitry. There are a number of techniques for improving their suitability, though, and they should be explored with the vendor. They could, however, involve significant engineering costs.

Mixed-signal Process Considerations

Next to full-custom (which is far too expensive and time consuming for most applications), cell-based design methodologies are the most flexible and offer a wide range of performance and accuracy options. Many ASIC vendors offer cell-based libraries of such functions as op-amps, comparators, A/D and D/A converters, filters, and passive components, and some offer parameterized analog cell generators. Array-based analog and mixed-signal design, on the other hand, tends to be too limiting. The design may be constrained by the combination of active and passive components or transistor types a given array provides. Array-based libraries also tend to be somewhat limited. As it is, most cell-based libraries require some level of customization or modification to satisfy particular application requirements.

The desirable process characteristics for digital and analog circuits are also in conflict, as most CMOS processes are optimized for digital designs. Their line widths are designed to be fabricated as small as possible, whereas analog designs typically require large transistors and wider line widths, especially in applications that require operating voltages greater than 5 V. As process feature sizes shrink in order to accommodate greater densities, analog performance suffers. These processes are simply unable to handle the large currents and voltages required by many analog applications. Processes optimized for analog circuitry, on the other hand, defeat the integration and performance requirements of digital circuitry.

Processes with larger feature sizes (2μ and 3μ) and two layers of polysilicon are much better suited to supporting the dynamic ranges of analog circuitry. The use of double poly allows the fabrication of more stable capacitors and other analog components that don't vary much over voltage and temperature. Most digital processes feature multiple levels of metal interconnect, but only one layer of polysilicon. In these processes, capacitors must be formed by sandwiching the oxide dielectric between a layer of metal and a layer of poly. Because of the dissimilarities between the two materials, precise values, ratios, and matching are very difficult to achieve. They also tend to drift with voltage and temperature variations.

Also, capacitors or resistors with large absolute values are generally impractical in ASICs (no matter what process is used) because their values are directly proportional to the area they occupy. A resistor's sheet resistance, for example, is measured in ohms per square, providing a simple calculation for die area for a given value. Die area, of course, is tightly coupled to die cost. Furthermore, precision components are very difficult to realize on an ASIC. As a result, most designers will use ratioed elements with low-value active current sources in place of high-value resistors.

In spite of its shortcomings for many analog applications, CMOS is a very popular technology choice for mixed-signal designs as it handles many analog functions very well (although its linear performance may not be as good as that of bipolar devices). The choice will ultimately depend on the ratio of analog to digital circuits on-chip, the type of circuitry, their respective performance requirements, and production volumes.

The ASIC Specification

The specification should contain all the information necessary to design, produce, and test the ASIC. Ideally, the specification should be complete and frozen at the initiation of the design cycle. But the functional definition of an ASIC is complex, and in most cases the design will start with an incomplete description and will evolve as the development cycle proceeds. The design cycle itself is often prolonged because of frequent changes in the specification. However, to make the most appropriate technology, methodology, design tool, and vendor decisions, the specification must be well defined. Keeping changes to a minimum will also greatly enhance the productivity of everyone involved in the design project.

Errors are most likely to occur as a result of specification miscommunications. Most often, such interpretation errors result from ambiguities in the specification. The circuit areas at greatest risk tend to be control logic and state machines. The design reviews take on greater importance in this sense. Reviews, therefore, should not be limited to the formal milestones. Rather, the design audit should be an ongoing activity. If specification changes occur, the design review process must be reset to the beginning so that the effects of design changes on the circuit (and the system) are fully comprehended. Both the design and the design tools must be able to accommodate design changes easily, particularly if the specification is likely to change.

Design Guidelines

ASIC vendors can most readily help avoid some of the more common pitfalls in ASIC design if they are involved early in the design cycle and continue to work closely with the design team throughout the design process. In addition to their assistance, the following considerations can help in every area of the design project.

I/O Considerations

• Most CMOS ASICs are designed with relatively low dc output drive, typically 4 mA. Some vendors offer I/O cells with greater drive capability (8 mA to 24 mA), but many times the lower drive pads must be combined in parallel to increase output drive. This approach can quickly consume output pads, however, if more than a few higher-drive outputs are needed.

• Provide an adequate number of power and ground pins. One rule of thumb is to allow one pair of power and ground pins for every eight outputs that might drive or change states simultaneously.

• Power and ground pins should be decoupled from the PCB via capacitors to provide some insulation from supply spikes. Designs typically needing extra power and ground pins include those that are large, have many outputs, have very high performance requirements, and have high output drive.

Simulation Considerations

Keep in mind that simulation serves two purposes: verifying circuit operation and developing patterns for test generation. These two objectives must be comprehended at the beginning of the design cycle if they are to be satisfied. The following guidelines will assist in this effort:

• Define all the simulation scenarios before beginning test vector generation. Have other members of the system design team (software engineers and board designers) review the simulation plan to ensure that no scenarios have been missed. How much simulation is enough? A rule of thumb is two to four test vectors per gate, but this is more a function of the circuit design rather than sheer gate count. Some circuits containing only 2,500 gates may require as many simulation/test patterns as some 10,000-gate circuits.

• Simulation should be performed at every level of the design hierarchy.

• The simulation should run without any errors or warnings. Every simulator error and unknown signal state should be tracked, understood, and fixed.

• Simulation should include off-chip loading. Output pins should be simulated with tester loading, as well as the conditions under which the ASIC will operate in the system.

• The simulation should always start with an initialization routine.

• Each set of simulation or test vectors should be self-contained and should not rely on pre-existing states resulting from previous vector sets.

• Develop test vector sets that can be compressed down to the maximum vector file size for a given tester. For example, most ASIC vendors allow a pattern depth of 4,096 vectors per file (Figure 7.1). Although this may sound restrictive, multiple files can be connected. Also, keep in mind that subroutines can be called and repeated and loops can be written as well. The effective number of vectors that may be applied with a single vector file, then, can potentially be many times the 4K limit for a single file. (ASIC vendors limit vector file sizes and total number of files not so much because of tester memory limitations—although available disk space is often an issue, but rather to limit the man-hours needed to load, debug and compile each file. The maximum allowable number of test vectors should, therefore, be a subject of negotiation.)

Basic Design Considerations

Even the most experienced designers of electronic systems using standard components must learn and utilize a new set of design guidelines when designing an ASIC. There are profound differences between designing at the board level and designing ASICs. The effects of placement and routing on performance, for example, are substantial. Also, with ASICs, it is often impossible to measure point-to-point delays inside the chip, whereas it is easy to probe pins on a PCB. Finally, rework of the ASIC, for whatever reason, can be extremely costly and time consuming; board fixes are relatively easy to fix, once the problem areas have been isolated. The

Location

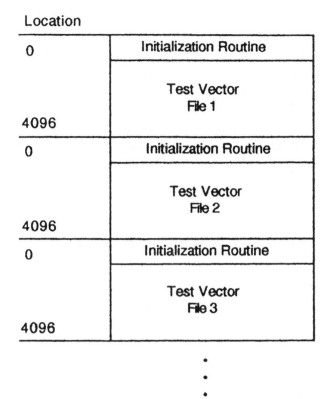

Figure 7.1 Test vector file representation.

following items should assist the engineer in addressing the primary areas responsible for creating difficulties when they are violated or ignored:

• Make effective use of the field applications engineer. Many aspects of the design process may not be documented. They can also help prevent problems from occurring by providing guidance that may be specific to the vendor's processes.

• Whenever possible, it is best to use synchronous design techniques. These techniques are less prone to be affected by delay skews. Wherever asynchronous signals will be introduced, a detailed SPICE simulation may be required to ensure that the circuit will operate correctly. Asynchronous inputs can result in signals with indeterminate states. All input signals should, therefore, be synchronized on- or off-chip. Synchronous design practice also eases verification, because all synchronous design elements run under a common clock. The idea behind synchronous design is that data is latched only after logic has settled (between clock periods). Remember that timing problems are extremely difficult to fix on an ASIC. If the circuit timing is marginal, the ASIC may be useless.

• Avoid excessive fan-out. If a node must be connected to many inputs, the preferred method is to buffer the node with either a large, high-drive buffer or a buffer tree. Large loads (common clock signals, for example) should be driven by a tree of buffers. All branches of the clock tree should be balanced to avoid skew. The use of large buffers will, however, increase the chip's power dissipation.

• Avoid excessive loading to minimize load-dependent rise and fall and signal delays. This can occur when an output is loaded by certain cells that have high input capacitances, such as high-drive buffers, I/O pads, set/reset lines on flip-flops, and XOR gates.

• Do not rely on gates as delay elements. Should the design later be migrated to another process, the delay-dependent timing will no longer be valid.

• Do not overdrive any one wire with multiple buffers. This condition can lead to metal migration and transients.

• Avoid gating together clock and data or control signals (Figure 7.2). Such circuits can best be described as *glitch generators*.

• Do not leave tristate buses floating when not in use.

• Provide a means of resetting or initializing all nodes to a known state within as few clock cycles as possible. The easiest method is to use flip-flops having reset or set lines.

• RAM signals should be accessible from the I/O so that standard RAM tests can be performed, thus easing test program development and debugging efforts.

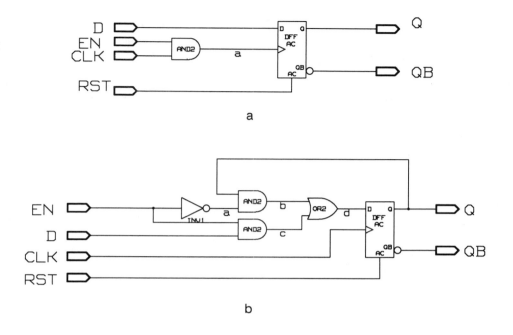

Figure 7.2 Asynchronous circuit including gated clock (a), and fully synchronized version eliminating the gated clock (b).

• Feedback loops should be latched. This will ensure that data on the feedback loop has settled before it is used. It will also prevent glitches from propagating through the circuitry, as well as prevent the circuit from going into oscillation.

• Many problems are related to data and clock skew, which can be caused by a poor timing specification, poor clock distribution, or poor data path distribution. The use of clock trees is recommended to avoid routing a clock line over a long distance. Minimizing clock fan-out to no more than four is also recommended to help localize clock problems, should they arise. Internally generated clocks should also be avoided, if possible, as they can cause glitches.

Design for Reliability

Design for reliability encompasses many disciplines, including logic design, physical design, test generation, device screening, and package reliability. A number of design techniques can be used to improve ASIC reliability. They include the following:

• Minimize the number of manufacturing defects through design-for-test techniques (see *Design for Testability*).

• Reduce the amplitude and duration of transients by using multiple power supply lines and separate power lines for the core area power distribution and for the I/O pad ring. Also, additional power and ground pads can be used to minimize switching transients and shorten the power paths to circuit nodes.

• Minimize the number of chip areas with high current densities by partitioning the layout into quadrants, each of which is supplied by its own power and ground ports. Power line widths should be sized according to the amount of current they carry.

• Provide adequate current drive in both the I/O and core circuitry, particularly for those elements driving large capacitive loads.

• Overdriving a node can result in large voltage transients. Rather than using a single large buffer to drive several loads, use a buffer tree composed of smaller buffers connected in parallel.

Design for Testability

Many design engineers would rather not become bogged down with the issues of test development. However, design engineers are in the best possible position to develop the ASIC test because they have the most knowledge of the design and can best evaluate how well the stimulus exercises the circuit.

Testing should therefore be an integral part of the design process, and it must be planned at the beginning of the design effort. After all, the test program is the ultimate specification for the chip. Failing to design without testability in mind will also translate directly into test development headaches later.

Increased levels of design integration have had the effect of restricting access to internal circuit nodes. Short of using expensive probing equipment, additional circuitry dedicated to easing access and testability must be included on chip. Testability can be enhanced through a number of methods. Circuit partitioning is a readily available alternative. Dividing the ASIC into smaller sections and multiplexing the control or observation points to the ASIC's I/O allows the manipulation and observation of nodes embedded within the circuit. Partitioning can be as simple as breaking a 16-bit counter, for example, into two 8-bit counters, each of which can be tested separately with significantly fewer vectors. To facilitate the testing of macro functions, cores, or other large blocks of logic, these elements can be isolated through the use of multiplexers. I/O pins may also be used as dedicated test points, allowing easy test access.

Of all the design-for-test methodologies, partitioning offers the lowest circuit overhead and has minimal impact on performance. However, it does not lend itself to automatic test pattern generation and may not increase testability to the desired level or simplify test development as much as other techniques. Other approaches to enhancing ASIC testability include scan test, boundary scan, and built-in-self-test, or BIST (a form of signature analysis).

Tester Considerations

Test program development is still a major bottleneck in the delivery of adequately tested prototypes. The alternative to prototype testing is no test at all. Devices shipped under these conditions are called "cut-and-gos." The shortcomings of this approach should be obvious, but may sometimes be necessary to get an early test run in the system while waiting for the test program to be developed.

125

ASIC prototype tests are typically derived from the simulation runs performed in the design phase. These simulation patterns are then translated or formatted for use on a given tester. Correlation between the resulting test program and the simulation patterns can be a tedious and time-consuming process. Problems commonly encountered in the test development phase include test engineers who know little or nothing about the device's functionality, the presence of uninitializable circuitry, noise, clock and signal skew, faulty load boards, and other problems that may result from poor design practices. There are tools available, however, that automate the translation process. These tools flag code violations and indicate how the patterns can be massaged into the correct format for the target tester. Alternatively, simulation patterns can be preformatted for a specific tester during the design phase.

The design and simulation plan must comprehend the specific device testers that will be used for prototype and production testing. For example, if scan techniques are to be used in a design, the tester must be capable of supporting them. That is, it must be able to store, scan, and compare the long strings of serial scan data. Some test equipment may not support scan at all, while others may require special interface hardware to support such tests. Understanding tester requirements and restrictions and designing accordingly will ensure a smooth transition from simulation patterns to an effective test program. The following guidelines highlight a few of the major areas for consideration:

• Simulations should be performed not only at the system clock frequency, but at the appropriate tester frequency as well.

• The simulation must be tester-oriented, that is, all simulation input timing must obey the synchronous, cycle-oriented timing constraints of the tester.

• Input stimulus should be applied at a rate that allows the outputs to stabilize before the next change in stimulus.

• Outputs should be strobed only when the circuit is completely stable.

• Specify test modes for more effective testing of large logic blocks or macro functions.

• Allow only one transition per clock period for any control or bidirectional node to avoid contention.

• Be mindful of setting up conditions that would result in pattern-induced timing hazards.

• Remember that the simulator cannot simulate switching noise. Therefore, it is critical to adhere to the design practices that minimize these effects so that they do not complicate test development or lead to operational failure.

• Break up long counter chains to reduce the number of clock cycles required to test their functions.

• Partition functions into logical blocks that can be isolated from the rest of the chip.

• Use initialization routines to reset or load circuit blocks with known values prior to testing.

• Use device test modes, where it is practical to do so, to make testing easier and more efficient.

Improving Test Coverage

It was discussed earlier that simulation is performed for two primary reasons: functional verification and test development. Likewise, simulation for test development serves two different purposes. First, the test program sorts the good parts from the bad. Functional and parametric tests ensure that the device performs within its specification. A second class of tests ensure that functionally good devices do not contain hidden manufacturing defects. Functional tests are typically only able to detect up to 70% of these potential manufacturing defects or faults. Fault coverage refers to the percentage of manufacturing defects that can be detected by the test program. The percentage of die on a processed wafer that might contain one or more faults is a function of the process maturity and cleanliness of the processing equipment and environment. The number of good die per wafer is referred to as the yield. The yield is a function of the die size (the larger the die, the fewer die per wafer and the greater the chance for a defect to be present) and the process defect density

127

(the number of defects statistically present in any square centimeter area of the wafer).

To determine the number of defects that might go undetected at any device yield and rate of fault coverage, simply multiply the average number of defects per die (should be less than one!) by (1 - fault coverage). For example, if the process produces an average of 0.80 defects per die and the fault coverage is determined to be 70%, 0.240 defects per die will go undetected. If the fault coverage were improved to 95%, .040 defects per die would go undetected. The increased level of fault coverage will necessarily lower the device yield, because more defects will be detected, resulting in more devices failing the test. Another method of evaluating the field failure rate is the Wadsack model:

$$(1 - Y)(1 - F)$$

where Y is the probe yield and F is the level of fault coverage. Thus, if the yield is 60% and the fault coverage is 95%, the field failure rate will be 2%.

Reducing the number of defects that might escape the test can improve system reliability and reduce the costs associated with field failures. Obviously, as the number of ASICs with low fault coverage per board increases, the chances for a field failure rise. These costs can be estimated by multiplying the number of undetected faults for each ASIC by the number of ASICs per board by the number of boards produced annually by the average field repair cost. Such an analysis can quickly point out the need for raising fault detection to as high a percentage as possible.

If the prototypes pass their functional tests, but fail in the system, the design could be wrong, or the test may be inadequate because it doesn't detect all the faults. The test strategy, therefore, should be used to assist in prototype validation as well. Debugging the chip is much easier when its designers have developed a comprehensive test plan.

There are several methods of improving an ASIC's testability and fault coverage (Figures 7.3 and 7.4). However, as circuit complexity increases, so do test development and associated circuit overhead. Because it is extremely difficult to generate an adequate test vector set after design completion, it must be integrated within the design cycle itself.

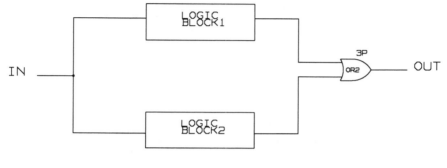

UNTESTABLE

a

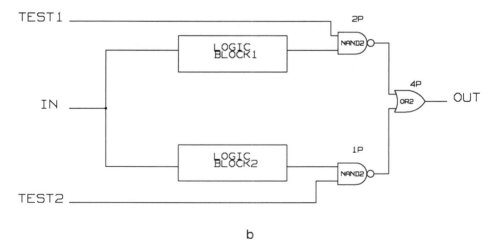

TESTABLE

b

Figure 7.3 Untestable circuit (a). Circuit corrected with test access (b).

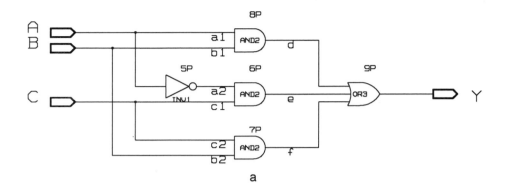

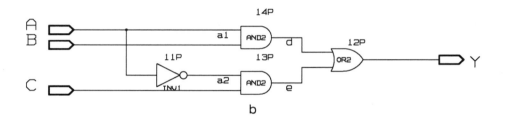

Figure 7.4 Redundant circuit with untestable nodes at b2, c2 and f (a). The circuit is made fully testable by reducing the logic (b).

Fault Grading

Although fault simulation measures and grades the fault coverage of the test program, its real goal is to help designers *improve* it. The U.S. Department of Defense requires, for many of its applications, that fault simulation analysis results accompany their products. They may also require a minimum acceptable level of fault coverage (95% or greater).

Fault grading provides a measure of how well a set of test patterns detect possible manufacturing defects. Fault grading tools use a variety of techniques to determine fault coverage (each with their own trade-offs) and provide diagnostic reports listing those nodes with undetected faults, areas of limited testability and the overall level of test coverage (expressed as a percentage of all circuit nodes that can be toggled by the test patterns). They are also used to track down the reason a fault does or does not propagate to a primary output. They may also suggest methods for improving fault coverage. This can be done either by changing the test program or by modifying the circuitry so that the fault can be detected. For these reasons, the fault simulator should be tightly linked to the logic simulator.

There are three primary fault grading techniques: probabilistic, deterministic, and statistical. Probabilistic fault grading (PFG) provides an estimate of fault coverage based on its evaluation of the design's connectivity and nodal activity (per the simulation). PFG tools are comparatively low in cost, but operate very fast. Their diagnostic capabilities are fairly good, but their accuracy is limited because the fault grading is estimated, not measured. Still, they are an excellent productivity tool that can help address most faults before submitting the design to the more expensive and time consuming deterministic fault simulation for final analysis and verification.

Deterministic fault grading (DFG) is extremely accurate because it compares the results of a good simulation with the simulation of the design with faults intentionally introduced into the circuitry. These faults are modeled as circuit nodes that are stuck at logic high or low levels. This methodology requires multiple simulation passes, each with a different fault, and therefore requires significant computing resources (accelerators) to be feasible. Some of the more sophisticated tools are able to display the results of the fault analysis graphically on the schematic, highlighting the nodes where the undetected faults are located.

The cost of achieving 100% fault coverage must be weighed against the probability as well as the cost of faults surfacing in the field. Therefore, an appropriate fault detection goal will depend on the application and the resources available to attain it. If an estimate of fault coverage is all that is required, then PFG will suffice. If, on the other hand, a high degree of accuracy and coverage is required, then DFG will be mandated. It is possible to combine PFG and DFG techniques and enjoy some of the benefits of both. For example, by applying deterministic fault simulation with only a small number of faults, a statistical measure of coverage can be attained and applied to the entire circuit. This method is known as statistical fault grading (SFG). Diagnostic information, however, will be somewhat limited.

It should be pointed out that current fault grading methods are not without their limitations. For example, the "stuck fault" model does not always correlate to actual faults or defects in silicon. The model is also inadequate for detecting faults in analog circuitry. Still, despite its shortcomings, a high degree of fault coverage *does* correlate well to high product quality and reliability.

Scan Test

The primary objective of design-for-test (DFT) techniques is to gain access to internal circuit nodes. By enhancing the ability to force a given node to a specified state (control) and evaluate the state of that node (observation), the level of fault detection is inherently increased.

There are many variants of scan testing, including level sensitive scan design (LSSD), scan path, random access scan and synchronous scan design (SSD). Although each implementation has its own clocking, data-latching, and accessing scheme, the basic idea is essentially the same and all greatly simplify the generation of test patterns providing complete fault coverage.

Scan design techniques are extremely effective in raising the level of testability, but are very costly in terms of added circuitry and resulting silicon area overhead, ranging from 10% to 20% (Figure 7.5). This is due to the larger size of the scan flip-flop, as compared to the area of a typical unscanned flip-flop, as well as the additional routing required for the scan path. Scan techniques also impose design constraints, which include requirements for fully synchronous design and no feedback loops in combinational logic. Circuit performance is also compromised due to the additional delay of the muxes that appear at each scanned flip-flop input.

132

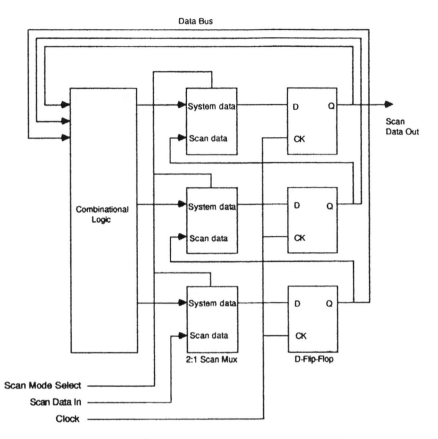

Figure 7.5 Scan test muxes inserted in D flip-flop path.

Because scan techniques tend to increase circuit area and also degrade its performance, many designers are opting to employ *partial* scan methods. This technique can provide high fault coverage, but does so with minimal design intrusion. The designer can scan only those circuit areas that require testability improvement.

In the basic scan path methodology, the circuit is designed so that it has two modes of operation: the normal functional mode and a test mode in which the circuit flip-flops are connected to form a shift register. When the circuit is in its test mode, an arbitrary test pattern can be shifted into the flip-flops. The circuit is then returned to its normal mode of operation for one clock cycle so that the combinational circuitry can act on the preset flip-flop contents (as well as primary input signals) and then store the results of its operations in the flip-flops. Returning the circuit to its test

133

mode, the contents of the flip-flops can be shifted out and compared with the expected or desired response. Scan test methodology does require, however, additional input and output pins for scannining data in and out of the device, as well as a test mode control pin.

Several suppliers offer tools that are able to automatically swap scan flip-flops for all the standard flip-flops in the design, daisy chain them together and generate test patterns as well.

Be advised also, that some testers may not be equipped to handle the long strings of serial scan vectors. The ASIC vendor may also impose extra test charges to accommodate scan test techniques.

JTAG Boundary Scan

The Joint Test Action Group (JTAG) has proposed a standard whereby the device incorporates a scan path through the inputs and outputs around the periphery of the device. Since the scan path goes around the perimeter of the device, it is referred to as the *boundary scan path*. This technique is particularly useful in developing board-level tests as it allows the ASIC to be isolated from other devices on the board. Test chains can be scanned around and into the ASIC when internal scan paths are multiplexed with the boundary scan path.

Implementing the JTAG boundary scan method requires four dedicated test pins to support the serial scanning cells. These are the test data input (TDI), test data output (TDO), test mode select (TMS), and test clock (TCK). These four signals constitute the test access port (TAP). In addition, implementation requires special control logic and registers. The TAP allows the serial read-in of data and instructions through the TDI pin and serial readout of data through the TDO pin.

Automatic Test Pattern Generation

Automatic test pattern generation (ATPG) generally requires the use of scannable logic elements to be effective. Its primary use is for the development of vectors to cover manufacturing defects, thus raising the level of fault coverage. Like other DFT techniques, ATPG also imposes similar design constraints.

134

Test Generation for Memory

The most common memory faults are opens and shorts in bits of the array, pattern sensitivity that causes certain bits in the array to be affected by the contents of an adjacent cell, and faults affecting the address decode circuitry. Test programs for RAMs must test for all such conditions. Testing memory on an ASIC can be difficult because all its address and data lines may not be accessible from the ASIC's I/O pins. Also, ASICs are most commonly tested on general-purpose testers, which may not be equipped with special hardware for RAM test pattern generation.

Once access to the RAM is established, though, several methods are available for testing it. These include alternately writing ones and zeros to and reading from each memory address location on the array and tests based on marching ones and zeros through each memory location in ascending order.

Since memory structures are so dense, they are much more sensitive to manufacturing defects. Therefore, it is critical to develop a test plan specifically for the on-chip memory blocks. Wherever possible, the address and data lines should be directly accessible through the I/O pins (through multiplexing, if necessary) to ease their testing and debugging.

Testability Trade-offs

The extra test circuitry required by scan techniques increases chip area, power, I/O pins, and circuit path delays. The scan techniques also impose design methodology restrictions and can potentially lengthen the design capture phase. Serializing test patterns, if not automated, can be a long and tedious process. Board-level testing becomes quite complicated if the board contains a mix of scan and nonscan components.

On the other hand, testable circuits require fewer vectors, resulting in shorter simulation times and straightforward test plans. Scan techniques allow the engineer to easily isolate the potential source of an error, thus minimizing the time needed for debugging and eliminating ambiguity in determining who is responsible for an error. Design for test also greatly enhances system reliability, resulting in significantly lower field repair costs.

Project Management

It is the project manager's responsibility to put together a design team that can satisfy all the design requirements and do so within the project budget and schedule. Responsibilities of all members of the design team, as well as the responsibilities of the respective vendors, must be well defined and understood. Team building is a critical element in meeting design goals. Proper consideration of the *human factors* can greatly enhance the chances for success. The requirements for a successful team include a competent team leader, well-defined project objectives, a program plan (whose development ideally includes team member participation), a committed, motivated team, and support from upper management. Each member must have a strong sense of ownership for his particular set of responsibilities.

The team must be able to perform the various design activities in parallel. Hierarchical design divided among several engineers is the best method for meeting tight development schedules. Each member of the design team must know all the design rules that must be followed, and all must have received proper design tool and methodology training. All communications regarding the specification and engineering changes must be written, formalized, and routed to all appropriate personnel. Although there are no substitutes for good designers equipped with the proper tools, effective project management can mean the difference between success and disaster for the ASIC design effort.

The Project Schedule

Ideally, the schedule is determined by working backward from the date that prototype units are required and arriving at a project start date. All too often, though, prototypes are needed "yesterday" and a more realistic method must therefore be used.

The selected design methodology will also have an impact on the schedule. Programmable logic-based designs, for example, can be developed in a matter of days, depending on the complexity, whereas some all-layer designs may take more than six months to complete. Highly leveraged, cell-based design methodologies, such as silicon compilation, can significantly reduce design and simulation times, but add time in the back end (mask and fab) due to their all-layer nature. Array-based designs, on the other hand, tend to treat all designs as flat (without hierarchy), which can introduce difficulties in design, simulation, layout, and verification. As a design methodology, it is inherently less productive. The back end activities for array-based designs, however, proceed more quickly, and iterations are less costly and time consuming. Time-to-market considerations should therefore be a major criteria in both the selection of a design methodology and the development of a project schedule.

The major events in the design cycle, which must be accounted for in the schedule, include the following:

- Specification development

- RFP preparation

- Vendor selection

- Design tool and library acquisition

- Logic design

- Design analysis (functional and preliminary timing simulation and analysis)

- Physical design

- Design analysis (final timing simulation)

- Design verification

- Prototype fabrication (mask, fab, package assembly, and test)

- Prototype evaluation and approval

- Production test program development

- Initiation of production

Schedules for each of the design reviews must also be included.

In addition to this list, if any special engineering or tooling is required, it must be included in the schedule as well. Each activity will require a block of time and some events may require iteration. Events that can cause schedule slippage include design and layout errors, design changes, design tool bugs, difficulties in design tool integration, and failure to perform by either party. Activities that are generally underestimated include simulation and design debugging. Lead times for packages, test program, and fixture development, as well as prototype fabrication, can become extended as well. Unexpected problems can also crop up that can have a significant impact on the schedule. For example, if the die size is larger than expected, it may require special lead frame or cavity tooling, which can take weeks or months to complete if no alternative tooling exists.

The schedule should be prepared according to typical and worst-case scenarios. Remember, also, that some ASIC vendors tend to be somewhat optimistic when quoting schedules. If the typical schedule were doubled, its impact on the project must be carefully evaluated. Time-to-market requirements should be able to tolerate such eventualities, as they are commonplace occurrences. Careful planning and management, though, can serve to abate most schedule risks.

Many vendors offer a processing throughput acceleration service that can take weeks off the development schedule. The services can be used up front or to make up for lost time. Such acceleration services, however, can cost $20,000 or more. Remember though, that haste makes waste, and that care should be taken so that extra money is not being spent to accelerate the delivery of nonfunctional devices.

It is important to keep the vendor apprised of the design progress and to schedule reviews and events as long in advance as is practical. This will ensure having a favorable position in the vendor's queue of incoming jobs, which can help preserve original schedule estimates.

Design Reviews

Design reviews are a critical project management activity in the design cycle. They occur at the primary milestones and are used to identify, correct, and prevent errors in the specification, design, and specific implementation. Depending on the expertise of the vendor, they may also be used to help optimize the design for performance, area, and cost.

A review should be held at each major design milestone, as follows:

- Design initiation

- Design completion

- Design verification

- Prototype approval

The initial design review provides an introduction to the specification and the basic design approach. The purpose of this first review is to ensure that the design approach and test plan will yield an optimized result. This meeting will also ensure that the design gets started on the right foot. The review consists of a meeting (at either the vendor or customer location) between the customer's designers, project managers, and other appropriate personnel and the vendor's design support engineers and project managers. A formal agenda should be drawn so that all necessary items for review are discussed and agreed upon. These items should include a detailed review of all written specifications, schematics, critical path timing requirements, design capture, simulation and test plans, as well as schedule and cost estimates. The interface between the two companies should also be determined and should include appropriate contact personnel, communications protocol, and reporting requirements.

Depending on the design methodology, the second review may occur after the completion of logic design and functional simulation, but prior to layout, or it may occur after the entire design has been completed. In the event of the review at design completion, it is assumed that the design is frozen and ready for final verification and to be committed to silicon. At this point, all the design information (including the design database and all appropriate specifications and documentation) is passed to the vendor for design verification. The design verification process, however, may uncover errors, which may require design, layout, or test vector changes, therefore potentially becoming an iterative process.

This should be the most rigorous of the reviews as it dissects the specific design implementation. It should ensure that correct design practices have been adhered to. It would include a review of the clock distribution, fan-outs, asynchronous circuits, potential bus contention problems, possible race conditions, and proper tristate bus design. It should also ensure that all design objectives have been satisfied, that is, critical path timing, output drive levels and loads, functionality, and performance under all conditions (best, nominal, and worst).

Upon completion of the first design verification pass, any resulting errors must be correlated, reconciled, and corrected. Once this process has been completed, a third design review is conducted whereby the design is signed off and deemed ready for fabrication.

Finally, a review is held to verify that the ASIC prototype passes all operational requirements per the test program and operates properly in the system. Should the device experience failures of any kind, a debug cycle is begun to isolate the problem(s) and corrective actions are implemented. The governing contract will assign liability for corrections, depending on their nature. In any case, both parties must cooperate and use their best efforts to resolve such potential difficulties.

The formal design reviews are absolutely critical in keeping the design project on track and minimizing the possibility for errors to become cast in silicon. The vendor will likely have a detailed checklist covering all items required for design submittal at each design stage (a copy of the design review schedule should be obtained as early as possible). Although reviews are held at formal milestones, they are by no means the only contact with the vendor. Design support from local and factory applications personnel should be ongoing throughout the design cycle. Their experiences with past ASIC verifications will undoubtedly have uncovered a number of problems that can easily be avoided.

The project manager's job is not complete until the ASIC(s) work in the system. Functionality at the board level is the final measure of success. In those situations where the chip does not perform to specification, it may be possible (if not preferable) to correct the problem through board-level fixes.

Time to Market Considerations

A new product's success is determined not only by its price/performance versus its competitors, but by its timely market introduction as well. Such windows of opportunity must be identified long in advance, and proper steps must be taken to ensure an adequate response. The developments in EDA have made it possible to rapidly turn product concepts into hardware; however, for many applications it is still not fast enough for comfort. Adding to the discomfort is the specter of design errors and iteration, which further delay market entry. Life cyles for electronics-based products have declined to about two years, and the window for market entry is generally no more than six months.

Sometimes design compromises must be made to meet introduction schedules. A design may require several iterations to achieve the full functionality desired, but as long as the product is among the first to market, some market foothold will have been achieved and such iterations can well be afforded.

In some cases it is possible to implement a design in programmable logic and thereby satisfy product introduction schedules quite easily. As the product ramps into production, the PLD-based design can be migrated to a mask-defined ASIC that can be produced far more cost effectively and may even include additional features.

Estimating the cost of being late or missing a market window all together is very straightforward. If, for example, a product should have generated $10 million if introduced on time into a market with an anticipated life of 18 months, a delay of 8 weeks (typical for a prototype iteration) would result in a revenue opportunity loss of more than $1 million. In addition, the loss of market leadership must also be counted should the competition get to market first. Even products with longer life cycles are not immune. It is extremely difficult to dethrone a market leader—even with a better mouse trap. All this amplifies the requirement for the ASIC, and ultimately the system, to be working right the first time, and on time.

Even with the most productive design tools, exhaustive design and system verification, the project is not out of the woods. Vendor misprocessing, accidental reversal of certain mask layers, misalignment, contamination, and errors in package wirebonding can still occur. An understanding of the fabrication, packaging, and test processes is therefore critical for the project manager in order to maintain accountability of the vendor for its quality control. This detailed understanding will also enable the development of a realistic project schedule.

Chapter 8

ASIC Prototyping and Verification

THE DESIGN OF AN ASIC that works right the first time on a tester load board is *not* the objective of the ASIC design project. Rather, the goal is to create a design that works in the system. ASIC vendors enjoy a fairly high degree of success in fabricating ASIC prototypes that work according to the test vectors. Unfortunately, that is no guarantee that the chip will work in the system. The reasons for such failures can almost always be attributed to an incomplete or misinterpreted ASIC specification or marginal timing. The need for system-level simulation, verification, and validation cannot be overstated. The vast majority of ASIC failures result from shortcuts or oversights in these areas. Every possible condition under which the chip will operate must be considered in its analysis. The essential question for the design team to ask is, "Are we convinced that this design is ready to commit to silicon?"

Design Verification

Once all parties are satisfied that the ASIC design is sound, the next task is to verify that the layout is good. That is the focus of the remainder of this section. Verification is the final step before fabrication begins. In this step, chip layout errors such as opens, shorts, and fan-out violations, as well as violations of design rule spacing requirements, will be caught. The following is a review of the verification checks typically performed by the vendor:

Electrical rules checks: The electrical rules checker (ERC) performs electrical connectivity checks on the circuit netlist, including short circuits, open circuits, and floating nodes. The user may also specify his own rules such as maximum fan-out and illegal circuit configurations. These types of checks are independent of any specific process technology, but electrical rules checkers also evaluate the design database for process technology-dependent errors, such as improper substrate bias and improper power and ground connections.

Design rules checks: The design rules checker (DRC) performs process design rule geometric spacing checks on the layout to ensure that the design artwork will be manufacturable in the specified process. These checks can be made on individual cells or on the entire chip layout database.

Layout versus schematic: Layout versus schematic (LVS) performs a comparison between the schematic netlist and its corresponding physical design database and reports any discrepancies. This guarantees that the mask layout matches the schematic connectivity.

Layout parameter extraction: This portion of the verification process extracts the parasitic resistance and capacitance of a node or net and outputs a SPICE netlist for use in a detailed analysis of critical path performance.

Prototype Fabrication

Fabrication is the point of no return for ASICs. Once the design is fabricated, errors are extremely costly in terms of both dollars and schedules. Depending on the device, methodology, and process, prototyping costs can range from $10,000 to more than $100,000 and require 6 to 18 weeks to complete. This underscores the need for thorough system-level simulation and exhaustive design verification before committing the design to silicon. Once the design is released, the wait begins—usually on pins and needles in anticipation of receiving the design in silicon. The following is an overview of the activities and methods used to build the prototype parts.

Mask Making

There are two basic approaches to exposing the wafer to the circuit patterns: projection alignment and direct write. The projection alignment approach involves a photolithographic process. The patterns are etched on transparent glass plates called masks or reticles. Each mask defines a subset of the circuit elements comprising the ASIC. The fabrication process may require as many as 10 to 24 masks to transfer the complete circuit image to the wafer, one layer atop another with overlay tolerances of 0.3 microns or better. In addition to tight alignment tolerances, the quality of the mask is critical. Any imperfections, defects, or dust particles present on the mask will be transferred to the wafer. To combat yield loss due to dirty masks, *pellicles*, a glassivation layer over the mask, stand off from the mask surface so that dust particles are defocused from the aligner's focal plane.

The circuit patterns are most commonly written to the mask plates via E-beam. The circuit patterns are generated by a data preparation program that takes as its input the design database, which includes descriptions of each circuit layer. The program then generates the patterns (output on magnetic tape) that are used to drive the E-beam machine used to make the masks. The tape is appropriately called a PG tape, for its pattern generation function. The most common data format for mask generation is GDSII. The same pattern generation program can also be used to generate plots of the circuit layout.

Projection alignment systems typically expose the entire wafer at once. Steppers, on the other hand, expose one die area at a time and step across the wafer to the next adjacent location, repeating this process until the entire wafer has been exposed. Although the stepper process is slower, it is capable of producing lines of finer resolution, thus better supporting fine-line processes.

Because of image field size limitations, the largest die size that can be supported using steppers is about 600 mils on a side. This limit is practical for yield considerations as well. The larger the die, the greater the chance it will be rendered inoperative due to manufacturing defects.

Another point of interest is that stepper reticles are typically made with two or three image fields. That is, two or three copies of the circuit pattern can appear on the reticle. Multiple fields are used primarily for quality assurance purposes. If at least one of the fields passes its QA inspection, the reticles will be usable. The fields, though, can be subdivided, thus allowing patterns for multiple ASICs to be combined on a single reticle set. Because of the field size limitations, however, the number of different ASIC designs that can fit into a single field is limited by the die size. Although additional reticle field inspection costs will be levied, the savings realized from sharing mask and wafer fab lot charges by several ASIC designs are substantial. Keep in mind, though, that while multiproject wafers (Figure 8.1) provide a savings

144

in prototype development, if large production volumes are anticipated, a new dedicated mask set will be required. Depending on the volumes, the charges for the mask set may be amortized with little impact on the production unit pricing.

Wafer Fabrication

The masks carrying the circuit images are positioned over the photoresist-coated wafer. Ultraviolet rays pass through the transparent areas of the mask, exposing the light-sensitive photoresist. Depending on the type of photoresist used, the exposed areas will remain or be removed in subsequent chemical etching steps, leaving behind the circuit image carried by the mask. As the process continues, additional mask layers further define circuit patterns and intermediate processing steps form transistor regions, insulating layers, vias, and metal interconnects.

Figure 8.1 Multiproject wafer. (Photograph courtesy of Orbit Semiconductor)

The light-sensitive photoresist material used to define circuit patterns on the wafer is also sensitive to electron beams. This gives rise to the ability to expose the photoresist directly via E-beam, bypassing the need for masks altogether. Instead of writing the circuit patterns on masks, they can be written directly on the wafer itself. Although the process is slow, the job can be done much quicker than the time needed to generate masks, and it does so at substantially lower costs. However, if production quantities are required, conventional masks are far more cost effective. The direct-write E-beam methodology makes possible very fast prototype turnaround. It also has the advantage of enabling layout error correction very quickly.

In addition, laser beams can be used to deposit the metal interconnect layers. The process uses tungsten metal instead of the more common aluminum, because at finer line widths tungsten is less susceptible to metal migration and flows in and out of vias better. Tungsten, however, has considerably more resistance than aluminum and, as a result, its performance is inferior to that of aluminum metallization.

Another alternative, best suited to arrays, is laser etching. Instead of depositing interconnect metalization, a laser is used to etch away unwanted metal, leaving only the desired interconnect patterns. Arrays specifically designed for this process have a solid sheet of metal covering the entire array area prior to personalization.

Wafer Probing

Wafer probing is a method of testing the individual die while they are still on the wafer. Because some prototype and preproduction packages can be more expensive than the die themselves, it is extremely useful to be able to identify the bad chips so that time and money isn't wasted assembling them. A wafer probe test is also required if the devices are to be delivered in die form for later assembly in a hybrid or chip-on-board.

A probe card (Figure 8.2) is essentially a PCB with metal traces that fan in to a position that will align with the chip's pad ring. The tests performed at wafer probe are generally dc parametric only. Once the chips are packaged, they will undergo more exhaustive parametric and functional testing. More extensive testing can be performed at the wafer level, particularly if the chips are to be supplied in die form, but the capacitive loading contributed by the probe card may limit its accuracy.

Wafer probing is conducted on dedicated equipment. The wafer is held in a vacuum chuck and positioned beneath the probe card so that the probe pins align with the pads on the die and make physical contact. Passivation openings at the bond

Figure 8.2 Controlled impedance probe card. (Photograph provided courtesy of Cerprobe)

pad sites permit the electrical connection. The test program is then run via cable connected to the tester. The probe station then automatically steps to the next adjacent die position, repeating this process until every die on the wafer has been tested. As each die is tested, the probe station will identify the die that fail the test by marking them with a red ink dot. The wafer is then scribed and sawn so that the individual die can be sorted (good from the bad) for packaging.

Depending on the circuit requirements and the cost of the package, wafer probing may not be performed during the prototype phase. Rather, a number of die, typically 40 from the lot of processed wafers, will be selected for packaging based on a visual inspection. The packaged units are then tested with the hope of finding at least 10 good samples that can be shipped to the customer. The cost of discarded packages must be weighed against the cost of the probe card tooling.

With the exception of gate array designs that have a fixed pad ring, each ASIC design will require a unique probe card tooling. Depending on the number and pitch of the pads to be probed, probe card costs can range from $500 to well over $10,000. Newer probe card technologies use thin-film cards in probing extremely high pad count, small pad pitch die.

Packaging

The probed wafers are then scribed and broken or sawn along the lines between the die and separated into individual chips. The good die are then most commonly assembled into ceramic packages that feature a removable cavity lid that facilitates debugging and inspection.

Prototype Test

Test program development has traditionally been the responsibility of test engineers. But as more complex ASICs are being designed, engineers, who generally lack test and measurement expertise, are discovering the need to assume many of the duties of the test engineer.

The importance of an effective test program cannot be overstated. The test program, after all, is the ultimate specification for the ASIC. It is the measure by which the device is deemed acceptable or unacceptable. In this sense, the burden is on the designer to develop the constraints by which the ASIC will be evaluated. If the ASIC fails in the field as a result of a fault that the test program hasn't detected or if the ASIC fails in the system, but passes its test program, it will be the user's responsibility to correct the problem. Because the device passes the test program, the vendor will be exonerated. In other cases, it may be more difficult to determine whether a problem is due to the device processing, the test methods, or the design. The test specification, therefore, should be as clear and concise as possible and well understood at the appropriate design reviews. It should detail those parameters that are 100% tested, parameters tested over temperature, loading conditions, input and output voltage levels, input and output leakage, pulse widths, timing delays, static and dynamic power consumption, and the specific conditions for each test.

The prototype verification process begins when the prototype samples are received from the vendor. The extent to which the prototypes have been tested by the vendor varies greatly, but, typically, their methods are inadequate to determine how well the part will operate in the system.

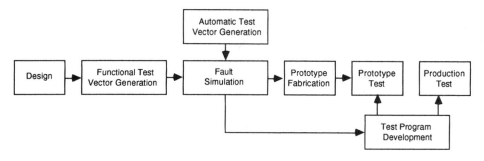

Figure 8.3 Test program development flow.

Test Program Development

As previously discussed, the simulator's output is not readily usable as a test program. This is due to the differences between the simulator's and the tester's speed, resolution, number of pins, amount of pattern memory, number of timing generators, and so on. While testers have limitations in all the aforementioned areas, simulators have no such limitations. Tester compatibility must therefore be built into the stimulus creation process. Assisting in developing test patterns and techniques are tools that evaluate the test patterns for compatibility with a given tester.

Tester rules checkers ensure that the simulation vector files conform to the target tester capabilities. They evaluate the vector file for such things as compatibility in data formats, timing generation, edge resolution, bus contention and I/O pin resources. The checker outputs a report detailing any violations, as well as an explanation and suggested remedy for each violation. At this point, the engineer can modify the design or the vectors to satisfy the constraints of the tester, or choose a more capable tester. If the rules check is okay, other tools are available to automatically generate a test program and test fixture wiring diagram.

If the test program is being developed to run on an ASIC verification system, the tester should have the capability to output test patterns in a format compatible with production testers, such as the Sentry series.

Tester Basics

The typical ASIC tester includes the following modules, which run under computer control:

• Time measurement unit (TMU): Consists of programmable clock generators, providing the clocking for the device under test (DUT) and controls input and output timing.

• Parametric measurement unit (PMU): Consists of a multimeter used for voltage, current, and resistance measurements and also provides several programmable voltage and current sources for the DUT.

• Test head: Provides the interface to the load board that carries the DUT.

• Load board: Configured or wired to interface with the DUT. Provides proper loading, supply voltages, and other I/O conditions.

• Memory and memory controller: The test patterns reside in the tester memory and are applied by the memory controller.

Testers have a number of test channels (each of which can be programmed as an input, output, or bidirectional channel) that are mapped to the pins of the DUT. The mapping is accomplished via load board wiring and software control. The number of channels and memory depth behind each channel that the tester can support varies considerably. Most testers are modular in nature, and their channel pin electronics and memory depth can be upgraded in increments.

Another restriction many testers have is the limited number of edge transitions that are allowed to occur within a specified period. Each period is referred to as a cycle and typically corresponds to a clock cycle for the ASIC under test. Because testers have a limited number of edge (timing) generators, the tester may be able to handle only a few transitions within each cycle.

Program statements force the voltages and currents to the DUT, thus setting up the conditions under which the device will be tested. During the test period, the appropriate signals (test vectors) are applied to each input pin (via the test channel), and the resulting output signals are strobed into the tester's comparator circuits. If the output patterns match the expected results, the device passes; otherwise, the part is rejected.

In addition to functional tests, dc and ac parameters must be measured as well. The dc parameters include supply, input load and output leakage currents, input and output high and low voltages, and so on. The ac measurements include pin-to-pin propagation delays and rise and fall times. Most vendors will permit the inclusion of several ac measurements (at no additional cost) to verify the timing of critical path signals.

At-speed Testing

Most ASIC devices are tested by the vendor at low speeds, typically 1 MHz. Although the commonly stated reason for low-speed prototype testing is that tester resources and capabilities are limited, the vendor would simply rather not perform testing at full speed because it tightens customer design acceptance criteria and increases the vendor's exposure.

Unfortunately, the fact that the ASIC might work at low speed is no guarantee that it will function properly at higher speeds. For example, when the chip is operated at its full speed, it will dissipate more power and heat up, leading to increased propagation delays. These effects will simply not be comprehended in a low-speed test. The effects of packaging on circuit parameters also become far more significant at higher operating frequencies. Critical path timing margins may also be tight, thus increasing the risk that the timing requirements will not be met. For these reasons, it is imperative that the tester support the full operating frequency of the device being tested. Full-speed testing should therefore be a primary acceptance criterion for prototype approval.

This situation is somewhat ironic because while so much attention has been paid to fine tuning the accuracy of the simulation, preserving that accuracy in the device test is an activity that has fallen short of the mark. It is simply foolish to assume that a device tested on a load board at 1 MHz will function in a system environment operating at 50 MHz.

Debugging the Test Setup

Many test problems are due to errors in the test fixture. Most commonly, the wiring diagram contains an error, or the fixture was miswired. The tester itself may be a source of testing errors if it is out of calibration. Also, the test load board may contribute leakage or voltage drops.

Noise is another common problem plaguing test setup. One method of isolating DUT switching noise is to switch one output, then two, and so on, until all outputs switch simultaneously or a failure results. On this note, it is also recommended to perform tests that exercise noise-sensitive circuits while injecting current surges and other noise sources into those circuits.

Characterization

ASIC vendors guarantee process parameters, but typically not the device—at least not to the extent that merchant market device suppliers characterize, qualify, and guarantee specific device performance and quality levels. These devices undergo extensive qualification and life testing of the die, the package, the die and package combination, as well as characterization of the device's dc and ac performance levels. Such qualification is generally not done for ASICs because each ASIC design is unique and only the design engineer has a complete understanding of the device's operation. Characterization, therefore, is difficult because neither the designer nor the ASIC vendor has all the information required to perform an exhaustive evaluation of the part. The designer typically does not enjoy access to the vendor's process characterization data, the distribution of best- and worst-case process parameters, and processing yield, because such data is considered to be sensitive and proprietary. In many cases, the vendor will not supply the data simply because it may not flatter the process. For newer processes, the data may not even exist. This is not to say that characterization of an ASIC device cannot be done. Arrangements for such work can be worked out, but at substantial cost (due to the extensive benchwork required). Furthermore, the vendor may not guarantee every characterization parameter.

Evaluating ASIC Testers

Not too long ago, ASIC testing was done using rack and stack test instruments such as logic analyzers, word generators, voltage and current sources, and multimeters. In response to the need for an integrated test environment, a number of testers designed specifically to support ASIC prototype verification, debugging, and characterization have emerged (Figure 8.4). These new systems combine the

Figure 8.4 ASIC verification system. (Photograph provided courtesy of Integrated Measurement Systems)

capabilities of the many bench instruments with the running of test programs derived from simulation vectors. Many also provide interfaces for wafer probers, device handlers, and data-logging equipment. The tester may also support debugging utilities such as acquisition triggering, pattern looping, and masking of bits or bit errors for particular pins. In addition, the following specific areas should be compared when evaluating ASIC verification systems:

- Maximum test rate

- Maximum test pattern depth

- Number of pins supported, cost per pin

- Links to simulator vectors

- Pin-to-pin skew

- Timing generator accuracy

- Edge placement resolution

- Compare strobe accuracy and resolution

- PMU accuracy and resolution

- Pin capacitance

- Driver output impedance

Each vendor's ASIC tester is unique, and their capabilities, as well as their limitations vary considerably. An effective evaluation of ASIC verification systems should be based not only on their actual test capabilities, but their integration with other EDA tools in the design environment as well. It should be pointed out also that ASIC test systems are large, expensive (if not outright overpriced) machines for which few industry standards exist. Software and data formats are generally tester-specific, and their transportation to other testers can be extremely difficult.

Chapter 9

ASIC Production Issues

IT IS ESTIMATED THAT the number of ASIC designs that survive to production is about 60% of all those started. Reasons for failing to reach the production stage include lack of market demand for the new product, late market entry, greater than anticipated costs (which render the product noncompetitive), and technical difficulties that cannot be reasonably resolved. For many of the products that do go into production, the production unit quantities may be far fewer than originally expected. In all these scenarios, the full costs of the engineering effort must be borne with little or no amortization over the production life of the product. This is one of the risk factors that must be carefully considered when embarking on an ASIC design project. Unless there is a large budget allocation for product research and development or fairly certain and justifiable production volumes, it might be well advised to consider initial product implementation with standard components or programmable logic. Assuming, though, that an ASIC design will go into production, the material that follows in this chapter will help to light the path to that destination.

The Preproduction Phase

Quite often, a number of prototype systems must be built to fully evaluate the system design. Several of the systems may be sent to potential customers for beta testing. Depending on the number of ASICs required to configure a system, the 10 prototype samples received from the ASIC vendor may not be sufficient. The minimum production order quantities that might be required by the vendor, on the other hand, may be overkill. In this in-between period, the preproduction requirements may well be satisfied from the material left over from the prototype wafer lot.

ASIC vendors will typically start from 10 to 24 wafers, depending on how they define a prototype lot. Regardless of the number started, not all wafers in a processed lot will be usable. One of the patterns stepped onto the wafer during processing is called a process control monitor (PCM). The PCM is a test die whose characteristics are probed and measured at various stages during the fabrication process and those wafers whose PCMs do not pass the parametric tests are rejected. Sometimes wafers are lost as a result of contamination or even butter-fingered operators. The wafer yield, though, is generally greater than 90%.

Since the typical number of prototype units delivered is 10, only one or more of the wafers will be used to satisfy prototype requirements. The remainder will be inventoried in anticipation of the first production requirements. It is important to understand that all those processed wafers from the prototype lot have been paid for by the user's NRE dollars (assuming the wafers do not contain multiple customer projects). Therefore, any preproduction quantities that can be satisfied by the remaining material should incur only wafer probe, package, assembly, and test costs to complete their processing. Depending on the die size, the yield from these wafers could be substantial.

Production Planning

Due to the lead times associated with each aspect of ASIC manufacturing (wafer fabrication, wafer probe, package assembly, test, and delivery), it is important to plan ahead for manufacturing requirements. Communication of material requirements to the ASIC vendor can be made in two forms. The obvious method is via order placement. The second method is through providing the vendor with a forecast of future material requirements. As the user demonstrates the accuracy of these forecasts (by converting them to orders), the vendor can justify building product to support the forecasted requirements. This in turn can then lead to vendor/customer programs such as just-in-time delivery (JIT), where product is delivered exactly when it is needed on the production floor. Such programs can be extremely valuable to the customer for their associated inventory management benefits and to the vendor for its own production planning purposes. Building to forecasted requirements as opposed to backlog does, however, involve some risk for the vendor. Since ASICs are custom devices, no one else can use them in the event the customer downsizes his forecasted requirements or cancels orders on the backlog. Likewise, users cannot liquidate their unused inventory of parts either. Also, pulling in deliveries of ASIC devices is not as flexible as it may be for standard components, because there is typically no box stock of the devices available. ASICs are built and shipped, for the most part, on an as-needed basis. If an unforecasted demand is made, a new lot must be started, and its delivery time will be subject to the associated lead times (although some vendors will support accelerated hot-lot processing). In such cases, gate arrays have the advantage over their cell-based relatives because the wafers are already preprocessed up to the point of metallization. The throughput time for processing two to four mask layers is weeks shorter than the time needed to process all layers. Be advised, though, that certain ASIC packages may have lead times longer than the wafer fabrication lead time. Tester capacity can also affect delivery schedules.

Quality Assurance

Quality assurance covers all manufacturing activities, including wafer fabrication, assembly, screening, quality conformance inspections, and documentation. An ASIC vendor's controls in these areas are generally documented in its quality manual, a copy of which should be obtained by the user's quality organization.

The required levels of ASIC quality and reliability depend on the application. For most commercial applications, testing consists of dc, functional, and ac measurements performed at room temperature. Certain industrial applications may require testing at an additional temperature, typically 85°C, and may also receive a static burn-in. ASICs designed for applications that require ultrahigh reliability (military, aerospace, and life support) receive far more rigorous testing. The applicable classes of quality can therefore be classified by the type of qualification and testing conducted.

Military-standard Quality and Reliability

DESC (Defense Electronic Supply Center) has established standards for IC quality and reliability and has published them in documents such as MIL-M-38510 (General Specification for Microcircuits) and MIL-STD-883 (Test Methods and Procedures for Microelectronics). These documents specify the qualification, screening, and inspection requirements for devices that are procured for military and other high-reliability applications.

157

The military standards define several levels of reliability that are applicable according to the type of device manufactured, as well as its operating environment. These include MIL-STD-883 (Classes B and S), JAN QPL, and QML.

MIL-STD-883, Class B: MIL-STD-883, Class B, Method 5005 specifies the device qualification and quality conformance inspections (QCI) performed on lot samples, while Method 5004 describes the 100% screening requirements for the devices. Each of these methods consist of subgroups that further define the specific tests and their configurations, sample sizes, and acceptance criteria. Per the standard, each unique ASIC design must be individually qualified.

The qualification and subsequent QCI flows consist of electrical testing (Group A), construction tests (Group B), die-related tests (Group C), and package tests (Group D). A description of the inspections in each of the groups is as follows:

- Group A (electrical tests)
    - DC (static) at 25°C, 125°C and -55°C
    - Functional at 25°C, 125°C and -55°C
    - AC (switching) at 25°C, 125°C and -55°C

- Group B (construction tests)
    - Physical dimensions
    - Resistance to solvents
    - Internal visual and mechanical
    - Bond strength
    - Die shear
    - Solderability
    - Hermetic seal

- Group C (die-related tests)
    - External visual
    - Temperature cycling
    - Constant acceleration
    - Hermetic seal
    - Steady-state life test
    - End-point electricals

- Group D (package tests)
  - Physical dimensions
  - Lead integrity
  - Thermal shock
  - Moisture resistance
  - Hermetic seal
  - Vibration
  - Constant acceleration
  - Salt atmosphere
  - Internal water vapor content
  - Adhesion of lead finish
  - Lid torque

As might be expected, substantial costs and lead times are associated with each QCI. The impact of both, however, can be reduced through a number of QCI strategies:

- Although Group A tests must be performed on every lot procured to the specification, a no-charge QA witness of the test setup can be substituted for a separate Group A inspection.

- The disposition of the Group C life test units is subject to the user's discretion. Should the units be considered to have been destroyed by the life test, then their costs (at the production unit price) are additive to the Group C inspection costs. If, on the other hand, they are considered to be worthy of production use, then only the Group C inspection costs will apply. Proper life test disposition is an issue that has long been debated. Some believe that units that have the survived life test have enhanced reliability, while others are of the opinion that the units have been overstressed and may be subject to early failure.

- Some Group D (package inspection) costs can be reduced if they are performed on electrical rejects.

- Groups C and D inspections, which are generally performed on a periodic basis (not required for each lot), can be replaced altogether through the acceptance of *generic* data, based on inspections of previous lots of similar material or evaluation vehicles gathered within a specified window. (The Group D generic data must be based on the same package type as that to be shipped.) Because sample quantities are required for the Groups C and D QCI (some of which are destroyed), significant cost and lead time savings can be realized through the acceptance of generic data. The vendor will, however, typically charge a nominal issuance fee, which can also be avoided if the vendor is instructed to retain the data on file, to be issued only in the event that the data is needed.

• In the event that the Group C life test must be performed, some costs can be saved if the burn-in boards can also be used for the life test.

Because ASICs are typically produced in small-lot quantities, the required QCI sample lot formation can represent a substantial percentage of the total lot cost. To relieve some of this burden, a relatively new method, 5010, has been approved. Method 5010 allows for smaller sample sizes and a reduced-scope QCI flow. Prior to the adoption of Method 5010, it was not unusual for the required inspection lot sample quantities to far exceed the number of units needed for the end application!

MIL-STD-883, Class S: Class S qualification and screening provides the highest level of device reliability. Applications for Class S ASICs include spacecraft, satellites, and other applications where failures cannot be tolerated or where repairs may be difficult or impossible to perform. Relative to Class B devices, building product to Class S standards involves more extensive screening and QCI procedures. Accordingly, lead times for Class S devices can be twice that of Class B and unit costs can be three to five times those for otherwise equivalent Class B devices.

100% Device Screening Requirements: Method 5004 defines the screening requirements that are to be performed on 100% of the devices in the production lot. Figure 9.1 provides a listing of the applicable tests for Class B and S flows.

Qualified Parts List (JAN QPL): To minimize the costs and lead times associated with JAN (Joint Army-Navy) or QPL (Qualified Parts List) ASICs, DESC has established qualification procedures that allow a blanket certification of a gate array

Procedure	Class B	Class S
Wafer lot acceptance	-	X
Nondestructive bond pull	-	X
Internal visual	X	X
Stabilization bake	X	X
Temperature cycling	X	X
Constant acceleration	X	X
Visual inspection	X	X
Particle impact noise detection	-	X
Serialization	-	X
Radiographic views	-	X
Pre-burn-in electrical parameters	X	X
Burn-in test (static)	X	X
Post-burn-in electrical parameters	X	X
Burn-in test (dynamic)	-	X
Post-burn-in electrical parameters	-	X
Final electrical test (dc, functional and ac at -55°C, +25°C and +125°C	X	X
Hermeticity seal	X	X
QCI sample selection	X	X
External visual	X	X

Figure 9.1 MIL-STD-883, Method 5004 100% screening tests.

family. Called the JAN Gate ArrayProgram, it provides qualification based on the generic (unpersonalized) base array family members. Qualification of the individual customer device personalizations (interconnect metalization layers) is not required under the program. Rather, the devices are built and screened (with Groups A and B inspections only) and shipped as JAN-qualified product. The customer simply details the design specifications in an altered item drawing (AID) for the qualified array. Otherwise, non-JAN ASIC designs must undergo extensive qualification procedures, requiring 6 to 9 months to complete, in order to qualify for listing on the QPL (full MIL-M-38510 compliance).

Qualified Manufacturer's List (QML): Because most ASICs for military applications are produced in small lots and the costs of qualifying each unique ASIC device is extremely high, it has been proposed that, rather than concentrating efforts on qualifying individual parts (per QPL), the focus of qualification should be shifted to the manufacturing processes themselves. This has given rise to the drafting of MIL-I-38535, also known as QML (qualified manufacturers list). Under QML, the certifying agency, DESC, conducts an exhaustive audit of the manufacturer's in-line process controls, design methodology (design tools and libraries), packaging, assembly, and testing operations. The qualification process uses standard evaluation and test chips, called qualification vehicles, rather than customer-specific ICs, to demonstrate control over all manufacturing steps and its compliance to strict QML requirements. Certification is granted when quality and reliability are shown to be the *result* of a manufacturer's methods and controls.

The net effect of working with QML-certified suppliers should be lower prices, shorter lead times, fewer documentation requirements, and faster access to new technologies for military ASICs.

Statistical Process Control

The traditional approach to quality assurance is to inspect the product after its manufacturing is completed. This end-of-line conformance verification is an inherently *reactive* approach. That is, quality is inspected in, not necessarily built in. Another drawback to this approach is the costly scrap that results. Conversely, an approach to quality that minimizes or prevents manufacturing defects through monitoring and correcting the mechanisms that can cause them is far superior. Statistical techniques can be used to provide insight into the various manufacturing processes, which can lead to corrective actions and continuous quality improvements.

Process variability has traditionally been a problem for manufacturers and users alike. It can result in parts that fail to operate to their specifications, as well as delayed deliveries due to yield loss at affected operations. Statistical process control (SPC) provides a means of quantifying, documenting, correcting, and improving process manufacturing performance. In addition to reducing variability, it increases process predictability and overall product reliability. The implementation of an SPC program should therefore be a requirement for the selected ASIC vendor, particularly if volume production is anticipated.

Electrostatic Discharge Controls

A familiar example of static discharge is the shock one receives from walking across a carpeted floor and then touching a doorknob. This shock is referred to as electrostatic discharge (ESD). ASICs, like most all integrated circuits, are static-sensitive devices. When ESD-induced voltages exceed the maximum signal levels, circuit operation will be upset and physical damage to the device may result. In such cases, the device is said to have been "zapped."

Due to the high cost of some ASICs (and their generally limited supply), few can afford losing devices to zapping. There are, however, several methods of controlling ESD and its effects. These include preventing it altogether, isolating circuit inputs from potential ESD sources, and increasing the ESD immunity of the devices themselves. Eliminating ESD completely is an almost impossible task. Air ionizers help to minimize the charge present in the handling environment, but, at best, they only serve to supplement other methods. These additional measures, which can help prevent static from ever reaching device inputs, include the following:

• The use of black conductive tubes has been shown to be more effective against static charge buildup than clear antistatic tubes.

• Once out of the tubes, the parts should be placed on black conductive foam.

• All handlers should wear grounding wrist straps.

• Conductive smocks should be worn.

• Polyester clothing (especially when worn with a wool sweater) should be prohibited in handling areas.

• Conductive floor tiles should be installed in handling areas.

• ESD handling precautions should be posted in all handling areas and all appropriate personnel trained in this discipline.

To make the device inherently less susceptible to ESD, diode protection networks are generally designed into the input pad circuitry. These circuits provide a path to ground through which the static charge can safely flow. The vendor should be able to provide ESD test data for its ASIC input devices, which, at a minimum, should be designed to survive ESD levels of 2,000 V (per the human body model).

Failure Analysis

The vendor should have a program in place to support material and failure analysis of nonconformant material. Special testers, curve tracers, logic analyzers, microprobers, SEMs (scanning electron microscopes), and methods supporting destructive physical analysis (deprocessing) are generally required to correlate device failures to physical defects on the ASIC. Some of the more common failure modes include contamination, broken or shorting wirebonds, opens and shorts in metal, and ESD damage. Figures 9.2, 9.3 and 9.4 illustrate a few of the more common failure mechanisms.

A failure analysis capability (whether in-house or subcontracted) is critical if the causes of such problems are to be discovered and appropriate corrective actions taken to prevent their recurrence. Requests for failure analysis are generally prompted by recurring anomalies which would seem to indicate a process problem. Infrequent, random or isolated device failures can generally be attributed to ESD.

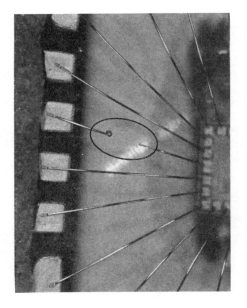

Figure 9.2 Fused-open bond wire, resulting in device failure.

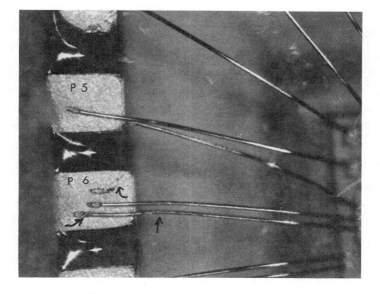

Figure 9.3 Pin 6 to pin 5 bond wire short.

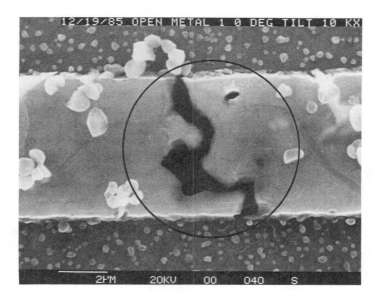

Figure 9.4 SEM photo of open metal.

Fab Capacity and Product Allocation

Unlike standard off-the-shelf components, ASICs are typically sole-sourced devices. This situation can pose serious supply jeopardy if the ASIC is produced in small quantities by a manufacturer who uses its excess capacity for foundry services. Those suppliers who provide merchant market semiconductors as well as ASICs quite often look to their ASIC manufacturing capacity for running standard products when demand for those products increases. To make room for standard part manufacturing requirements, suppliers may push out ASIC deliveries, increase prices, or both. The reason is that large-volume runs of standard components in an allocation market environment are far more profitable than small runs of individual ASIC designs.

Most major merchant market semiconductor houses use subcontracted foundry services as well, but typically only for commodity devices such as SRAMs and DRAMs. Most of this foundry work is performed offshore (Japan, Korea, and Taiwan), as is most package/assembly work (Philippines, Hong Kong, and Korea). Offshore subcontractors are able to manufacture at substantially lower costs than domestic sources due to the inexpensive labor in those areas. Semiconductor makers also like using foundries because they are able to ramp them up and down quickly in response to changes in demand, keeping their own wafer fab facilities operating evenly.

Alternate Sourcing

For an electronic product to maintain its competitiveness, it may be necessary that the design be migrated to a newer, more aggressive process. For example, if a design was originally fabricated in a 1.5μ process, as the product matures, competitive pressures on system speed and cost may require that the ASIC be redesigned in a submicron process. The extent of the redesign effort will depend on the design methodology originally used. Process-independent silicon compilers, for example, can automate the physical design (layout) conversion to another set of design rules, but will still require the design to be resimulated to ensure that timing violations will not occur in the faster process. The recompiled design will then have to go through the same prototyping phase as the original design.

Likewise, gate array and standard cell netlists can be mapped to cells in new or alternate libraries and processes. This is a potentially expedient method of alternate source development, however, in some cases, it may not be possible or may be more difficult than the original design effort. If, for example, the design contains high-level macro functions, they may appear as *black boxes* in the netlist. In these cases, flattened netlists of the macros (all design hierarchy removed) may not be made available by the original vendor, precluding mapping to a competitor's library.

The particular design practices employed in the original design may also make the design conversion difficult. If the design contains gated clocks, asynchronous signals, large fan-outs, unlatched feedback loops, or timing-specific delay elements in timing-sensitive paths, it may be impossible to achieve a clean simulation in the new library/process. This reinforces the need to adhere to strict design practices.

Chapter 10

ASIC Cost Determination

THE TOTAL COST OF AN ASIC must include the amortization of all nonrecurring engineering charges, including those for design, design tools, special tooling, prototyping, device evaluation, and debugging and iteration on any of these activities. The analysis of ASIC costs should also comprehend break-even costs as they relate to the equivalent design implementation in standard components.

ASIC Production Unit Cost Factors

In determining the production unit price for an ASIC design, a number of factors must first be known. These include costs that are primarily a function of the die size, process, and packaging used, as well as the level of screening performed.

Estimating the Die Size

Die size estimation is as much a function of the design as it is of the methodology and technology used to implement it. In the case of gate arrays, the various die sizes in the family are fixed; die size estimation is merely a matter of identifying the appropriate array for the design. This is generally based on equivalent gate count and I/O requirements. However, as has been previously discussed, the device architecture, as well as the design architecture and its associated routing requirements, can greatly complicate this simple measure. Cell-based designs require even greater accuracy in die size estimations. The following provides an overview of the issues involved in arriving at an accurate die size estimation that will provide the basis for design feasibility and cost estimating.

There are two primary elements in determining the die size: the core area and the I/O pad ring. The core size will be the primary yield-limiting factor, while the pad ring, in pad-limited cases, will be the be the factor limiting the number of candidate die per wafer.

Core area estimates can be developed in a number of ways. The simplest method is to add up the number of equivalent gates for all the functions in the design. Each gate occupies a certain, fixed area. The total gate area is then multiplied by a factor representing the routing overhead for the design. Depending on the design architecture, this factor can range from less than two (for dense data path and RAM-based ASICs) to as much as three or four times the total gate area (for random logic and designs containing several large global buses).

Compiled custom design can provide some advantages in die size estimation accuracy over simple, library-based methodologies. Because of its higher-level design capture approach, one can compile the constituent design elements rather quickly, connect the major buses, and arrive at a fairly accurate die size estimate. Regardless of the design methodology though, there is really no substitute for benchmarking the design for die size estimation.

The other die size factor is the I/O pad ring. A simple formula will determine whether the design is core or pad limited:

$$\frac{Pad\ Count\ *\ Pad\ Pitch}{4} + (2\ *\ Pad\ Height)$$

The pad count is the total number of pads the design will require, including those for power and ground. The pad pitch is the center-to-center spacing between two adjacent pads. Most wirebonders will accommodate a pad pitch of as little as 5.5 mils, while TAB methods will handle pad pitches of as little as 4 mils. In some cases, for grossly pad limited designs, it is possible to stagger the pad placement. That is, one pad ring is placed inside the outside pad ring and offset such that bond wires can be attached without shorting the wires of the outside ring. This arrangement can have a significant impact on die size as it essentially cuts the effective pad pitch in half. Note, however, that the pad height (pad plus pad driver circuitry) will be doubled. Assembly costs and yields may also be negatively affected and wafer probe card costs will likely be substantially higher. These costs, though, must be weighed against the potential savings realized from the smaller die size.

169

The area for the pads (number of pads * pad pitch) is then divided by 4 to provide the number of pads per die edge. The pad height for two opposite sides of the die is then added to this figure. Typical pad heights range from 12 to 18 mils, depending on the type of pad driver circuitry used. Finally, the distance to the scribe center line (typically 2 to 4 mils) must be included. The scribe line is the channel that separates individual die on the wafer and provides the area for wafer scribing or sawing. (Most of the preceding discussion is applicable only to cell-based designs because the number, locations, and sizes of pads on gate arrays are fixed.)

Differences among the various design methodologies will also affect the resulting die size. For example, compiled-custom designs typically achieve layouts that are far more dense than the equivalent implementation in a gate array. In fact, it is not uncommon for the area of a compiled-custom design to be as little as half that of the gate array version. There may also be significant differences between competing cell- and array-based methodologies. So much depends on the architecture, the layout of the primitives and macro cells, and the routing efficiency of the design tools.

The differences between processes have an equally dramatic effect on die size. As discussed in Chapter 3, there can even be a significant disparity between two processes that are both designated as 1μm. Also, the number of metal interconnect layers supported by the process will result in differences in routing efficiency and the ratio of routing area to active circuitry area. The conclusion one should rapidly come to is that for many applications there is no easy way to accurately predict what the final core and die sizes will be without performing some type of benchmark. Die size benchmarking should be considered a critical activity in the selection of the ASIC approach, because so much of a project's technical and economic feasibility depends on it. For designs that require larger die sizes, small incremental increases in area can have a substantial impact on the die cost. Once an estimate is made, the vendor should guarantee the limits. The increased cost of the chip, should it push those limits, should be well understood.

Predicting Process Yield

There are many disciplines involved in producing an ASIC with good yields. These include the areas of design, fabrication, assembly, and test. Depending on the specific manufacturing controls an organization has established, one group may contribute more than others in the effort. Others may make a greater contribution to fallout. Ongoing quality improvement in each design and manufacturing discipline should therefore be demonstrated by the vendor. This would normally include statistical process controls and well-trained and indoctrinated personnel.

Perhaps the most significant variable in determining ASIC cost is the wafer probe yield. Probe yield is simply the percentage of good die on a wafer. Because process yield is directly related to production unit costs, it is important for the designer to have a reasonable idea of how well a design will yield before it is actually built. Due to the erratic yields often encountered with chip sizes of 500+ mils/side, delivery dependability can easily become jeopardized. Probe yield also limits the level of integration possible with a given process.

Yield loss is due to manufacturing defects such as opens and shorts that are caused by particles, mask misalignment, mask quality, poor step coverage, photoresist debris, and other sources of contamination (many of which originate from the chemicals and the water used in the process and are typically visual in nature). The effective defect density, expressed as the number of defects per unit area (generally 1 cm^2), determines the manufacturing yield for a given process.

The defect density is a random element. It varies across the wafer, as well as from wafer to wafer and from lot to lot. As a result, yield variations from wafer to wafer and lot to lot can be significant. A fairly large sample size is therefore needed to fully calibrate the yield models. Unfortunately, process defect density information is not readily released by vendors. In fact, most vendors guard process yield information so closely that it will generally not be disclosed even under terms of a nondisclosure agreement. Rather, it typically must be derived from the die size and the probe yield information. The probe yield can only be known if the material is purchased in wafer form and probed in-house or through a subcontractor.

The defect density value is of particular importance because small, incremental changes can have a tremendous impact on the yield. For example, Figure 10.1 shows the yield in net die per wafer for a die measuring 300 mils/side by defect density and wafer size.

Defect Density	4 in	5 in	6 in
1	59	99	150
2	42	70	106
3	31	53	80

Figure 10.1 Effect of defect density on yield.

Figure 10.2 provides a representation of the same pattern of defects on two wafers, each containing a different die size, graphically highlighting the impact of both die size and defect density on yield. The wafer on the left will have a yield of approximately 50%, while the wafer on the right will yield about 90% of the candidate die.

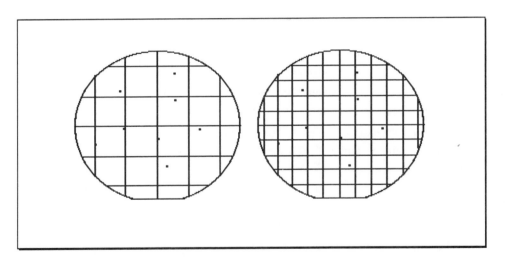

Figure 10.2 Defect density versus die size.

There is a substantial learning curve for ASIC manufacturing. This learning curve is complicated further by the small production runs typically associated with ASICs. The process flow must be stopped, started, and restarted for every unique production lot on the line. The learning curve particularly affects new processes, which may take one to two years to achieve even typical yields. With every doubling of the production volume, though, the cost should decrease 20% to 30%.

There are a number of statistical models for predicting process yield. All such models are generally based on the average number of defects per unit area (D_0) and the square area of the chip (A). Some models consider the density of the circuitry (RAM versus random logic), total utilization, as well as the number of critical mask layers in the process. For very small chip areas, all the models correlate extremely well. However, as chip size increases, the various models exhibit quite a disparity. The following is a review of the most commonly applied yield models.

The Poisson model assumes that the distribution of defects is constant across the wafer and from lot to lot. As a result, the model becomes increasingly inaccurate with increased die size. The Poisson model is:

$$Y = 1 - e^{-AD}$$

Murphy's yield model, on the other hand, comprehends that the defect distribution is not uniform and is given by:

$$Y = [(1 - e^{-AD}) / AD]^2$$

The Murphy model, based on an approximation of a bell-shaped Gaussian distribution, is accurate for smaller die sizes, but tends to be pessimistic for larger die sizes. The Seeds model, on the other hand, is based on an exponential defect density distribution and tends to be optimistic for larger die sizes.

$$Y = e^{-\sqrt{AD}}$$

In correlating these models to actual device yields, it has been observed that the results lie between the Murphy model and the Seeds model for yields below 33%. Above 33%, the Seeds model tends to be pessimistic, while Murphy's model may be more accurate. In any case, the two models generally define the upper and lower limits, thus an average of the two yields a prediction that is more accurate than either model used alone (Figure 10.3).

Once the anticipated yield of a given process is known or reasonably estimated, the number of good die that can be yielded from a single wafer (on average) can be estimated. The number of candidate die on a wafer (or gross die per wafer) is given by

$$\text{GDPW} = \frac{\pi(r - \sqrt{A})^2}{A}$$

where r is the radius of the wafer and A is the square area of the die. Multiplying gross die per wafer by the projected yield provides an estimate of the number of electrically good die per wafer—*according to the wafer probe test.* The area at the wafer's perimeter must be discounted because it is essentially useless. Also, PCMs (process control monitoring test die) further reduce the potential number of good die (there are typically four to six PCM locations on the wafer). Net die per wafer at wafer probe (NDPW), then, can be approximated by the following:

Murphy model:

$$\text{NDPW} = \frac{\pi(r - \sqrt{A})^2}{A} [(1 - e^{-AD}) / AD]^2$$

Seeds model:

$$\text{NDPW} = \frac{\pi(r - \sqrt{A})^2}{A} (e^{-\sqrt{AD}})$$

Murphy-Seeds average:

$$\text{NDPW} = \frac{\pi(r - \sqrt{A})^2}{2A} [((1 - e^{-AD}) / AD)^2 + (e^{-\sqrt{AD}})]$$

The impact that the die size and its corresponding probe yield have on die cost is shown in Figure 10.4. The chart assumes a 6 in. wafer at a processed cost of $600 and a defect density of 2.0.

The larger the die, the fewer die per wafer available, and the greater the chance that any die will be affected by a manufacturing defect. Also, the greater the circuit density, the greater the probability that a defect will affect circuit functionality. Likewise, there will always be areas on the chip where manufacturing defects will not cause a device failure. These would include areas of sparse logic and routing, the I/O pad ring, pad driver circuitry, and the moat route area (area dedicated to routing signals from the core to the I/O pad ring). Correction factors can be applied to the

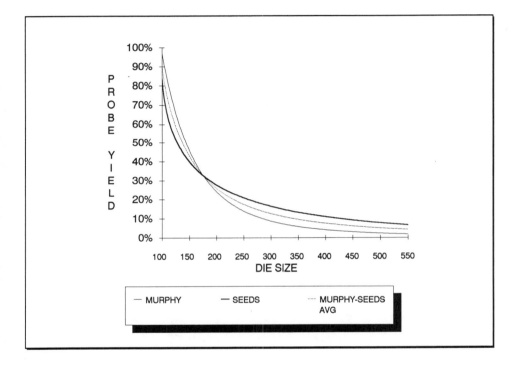

Figure 10.3 Probe yield as a function of die size.

nominal defect density to better approximate an *effective* defect density for a given chip. More accurately, the defect density can be modeled to affect only the core area of the chip (the area that contains the defect-sensitive active circuitry). No matter what the defect density though, a chip consisting primarily of random logic with a high degree of routing will yield far better than a chip of the same size consisting primarily of RAM.

The costs due to yield loss from operations including wafer scribe and break, assembly, final test, and quality inspections must also be factored into the analysis. The final yield is given by

$$Y_{final} = Y_{wafer} Y_{probe} Y_{scribe\ and\ break} Y_{assembly} Y_{final\ test} Y_{reliability\ testing}$$

Following the wafer probe operation, the wafers are scribed along streets between the die locations, broken apart, and sorted. Problems that can result from this operation include damage from cracks, chips and contamination. The typical yield for the procedure, though, is 95% or better.

The assembly yield is determined by the number of mechanically good packaged parts yielded out of assembly, divided by the number of good die submitted to the assembly operation. Depending on the package type, number of pins, and the pad pitch, the assembly yield can range from 80% to 98%. Most assembly defects (the source of the majority of reliability problems) are the result of broken or shorting bond wires, wirebonding errors, and broken or cracked die.

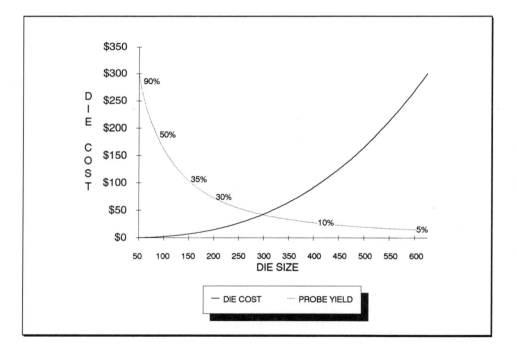

Figure 10.4 Die cost versus die size and probe yield.

The final test yield is determined by the number of electrically good packaged units yielded out of the final test, divided by the number of mechanically good packaged units submitted to the final test operation. Depending on the number of pins, circuit complexity, performance requirements and fault coverage, the final test yield can range from 80% to greater than 95%.

Reliability tests, including burn-in and other environmental screening tests, contribute further yield loss. Burn-in typically carries a percent defects allowable (PDA) of 4% to 5%.

Calculating the Production Unit Price

The production unit price is affected by the die cost, package and assembly costs, and final electrical test cost. The production order quantity also plays a large part in determining the final unit cost.

Factors affecting die cost are the die size, wafer size (4, 5, 6, or 8 in.), wafer cost, and process yield. The process yield is primarily a function of clean room effectiveness, process resolution, and maturity. The number of wafer starts per week also affects the vendor's ability to attain manufacturing economies, thus minimizing equipment amortization and overhead. To determine the die cost, first calculate the total number of wafers needed to yield a desired number of die (this number must include the number of die that will be lost due to yield fallout from assembly and test operations). Multiply this number by the wafer cost and divide the result by the total number of good die yielded. Add to this the cost of the wafer probe operation, which is calculated by multiplying the single wafer probe cost by the number of probed wafers, divided by the total number of good die.

Next, the package cost must be determined. For plastic packages, this cost will include the cost of the lead frame, die attach and wirebonding operations, the molding compound, and labor and equipment depreciation (which is higher for new toolings). Plastic assembly costs typically range from $0.01/lead to $0.04/lead, depending on the package type, quantity, and assembly location (offshore labor rates are substantially lower than domestic rates). Ceramic package costs include the base package, assembly costs, and the lid. For most military applications, assembly must be performed at certified state-side locations. The nature of the high-reliability assembly process combined with higher domestic labor rates results in much higher assembly costs (5 to 10 times greater). Also, production lots tend to be small, further increasing costs (due to lot charges that must be amortized over relatively few parts).

Once the die are assembled into the package, they are submitted to final electrical test. Like the assembly operation, the number of packaged parts submitted to the test operation must comprehend the final test yield (in part, a function of how thorough the wafer probe test is). The cost of the final electrical test depends primarily on the pin count (higher pin counts may require more expensive testers), the type of test performed (analog or digital), and the time required to perform the test. Testers are generally operated on a cost per hour basis. The time charged includes all setup and load times in addition to the actual test time. The test time is largely a function of number of test vectors and tester clock frequency. The per unit test cost can be determined by dividing the total tester hours cost by the number of devices tested. Most ASIC vendors will allow a test time of 2 seconds per device in their standard production unit pricing.

Finally, the selling price of the ASIC is determined by a variety of market and cost factors, which include the following:

• Sales

• General and administrative overhead

• Research and development

• Condition of the economy

• Wafer processing capacity and loading

• Competition

• Volume ordered

• Amortized NRE

• Desired gross profit margin (GPM)

The selling price is calculated by

$$\frac{\text{cost}}{1 - \text{desired profit margin}}$$

If, for example, the total cost of the ASIC is $10 and a 40% GPM is desired, the selling price would be $16.67.

Figure 10.5 provides a simplified cost calculation example for a die measuring 300 mils/side, packaged in a 84-pin PLCC, and delivered in a quantity of 10,000 units.

In some cases, the ASIC vendor will agree to amortize a substantial part of the prototype NRE over the first few production unit orders. Under such circumstances, these costs must be added to the unit price. Although such an arrangement has no impact on total expenditures, it does provide a means of manipulating internal budgets for design development versus production.

Processed wafer cost (6 in.)	$400
Wafer probe cost	$50
Probed wafer cost	$450
Gross die per wafer	256
Wafer probe yield	45%
Good die per wafer	115
Number of wafers needed to ship 10,000 pcs	90
Die cost	$4.05
Yielded package and assembly cost	$1.32
Yielded final test cost	$0.54
Total cost	$5.91
Selling price @ 30% GPM	$8.44

Figure 10.5 Production unit cost calculation.

APPENDIX A

ASIC Design Evaluation and Pricing Programs

BECAUSE THE VARIOUS ASIC YIELD AND COST calculations are complex and potentially confusing, the author provides an integrated set of PC-based programs that perform all calculations automatically. The programs allow the user to calculate actual detailed costs and market prices for virtually any ASIC configuration in a variety of processes, die sizes, packages and production volumes. The programs include updated wafer and package costs as well as defect density values. The CHIP•COST™ Program is the same tool the ASIC vendors use to price their customers' chips. This program will allow the user to generate his own quotes for comparison and calibration with the actual quotes that would be received from the candidate ASIC vendors. The program will also facilitate the evaluation of different partitioning schemes for an ASIC design, perform budget analysis, identify potential yield and delivery problems, and provide the information necessary to negotiate from a position of strength.

The package also includes programs for estimating die size, power dissipation, and thermal capabilities of most packages. Programs for determining ASIC development schedules and performing objective ASIC vendor evaluations are also provided.

To receive your copy of the integrated set of PC-based, menu-driven programs (including documentation), send $24.95 to:

ASIC Programs
PO Box 49241
Colorado Springs, CO 80949

APPENDIX B

Directory of ASIC Vendors

ABB HAFO Inc.
11501 Rancho Bernardo Rd.
San Diego, CA 92127
(619) 485-8200

AT&T Technologies, Inc.
555 Union Blvd.
Allentown, PA 18103
(800) 372-2447

Custom Arrays Corp.
525 Del Rey Ave.
Sunnyvale, CA 94086
(408) 749-1166

Actel Corp.
955 E. Arques Ave.
Sunnyvale, CA 94086
(408) 739-1010

Barvon BiCMOS Technology
1992 Tarob Ct.
Milpitas, CA 95035
(408) 262-8368

Custom Silicon, Inc.
600 Suffolk St.
Lowell, MA 01854
(508) 454-4600

Adams Russell Semiconductor
80 Cambridge St.
Burlington, MA 01803
(617) 273-5830

California Micro Devices Corp.
2000 W. 14th St.
Tempe, AZ 85281
(602) 921-4541

Data Linear Corp.
491 Fairview Way
Milpitas, CA 95035
(408) 945-9080

Advanced Micro Devices, Inc.
5900 E. Ben White Blvd.
Austin, TX 78741
(512) 462-5667

Cherry Semiconductor Corp.
2000 S. County Trail
East Greenwich, RI 02818
(401) 885-3600

Design Devices
20301 Century Blvd.
Germantown, MD 20874
(301) 428-6660

Applied Micro Circuits Corp.
6195 Lusk Blvd.
San Diego, CA 92121
(619) 450-9333

Coherent Design
275 Saratoga, Ste. 200
Santa Clara, CA 95050
(408) 296-3710

Design Engineering, Inc.
1900 13th St., Ste. 304
Boulder, CO 80302
(303) 440-7997

Arrow Electronics, Inc.
25 Hub Dr.
Melville, NY 11747
(516) 391-1300

Commodore Semiconductor
950 Rittenhouse Rd.
Norristown, PA 19403
(215) 666-2585

Electronic Technology
525 East Second St.
Ames, IA 50010
(515) 233-6360

ASIC Northwest, Inc.
13353 NE Bel-Red Road, Ste. 202
Bellevue, WA 98005
(206) 643-0919

Control Data Corp.
8100 34th Ave. So.
Minneapolis, MN 55440
(612) 853-3117

Exar Corp.
2222 Qume Dr.
San Jose, CA 95131
(408) 434-6400

Ford Microelectronics, Inc.
10340 State Hwy 84 North
Colorado Springs, CO 80918
(719) 528-7660

Hall-Mark Electronics Corp.
11333 Pagemill Rd.
Dallas, TX 75243
(214) 343-5923

Integrated Circuit Tech.
22691 Lambert St.
El Toro, CA 92630
(714) 581-7195

Fujitsu Microelectronics, Inc.
3545 North First St.
San Jose, CA 95134
(408) 922-9000

Hamilton/Avnet
1175 Bordeaux Dr.
Sunnyvale, CA 94089
(800) 426-2742

Integrated Logic Systems
4445 Northpark Dr.
Colorado Springs, CO
(719) 590-1588

Gain Electronics Corp.
22 Chubb Way
Somerville, NJ 08876
(201) 526-7111

Holt Integrated Circuits, Inc.
9351 Jeronimo Rd.
Irvine, CA 92718
(714) 859-8800

Intel Corp.
6501 West Chandler Blvd.
Chandler, AZ 85226
(602) 961-2801

GE Microelectronics Center
One Micron Drive.
RTP, NC 27709
(919) 549-3100

Honeywell Inc. (Atmel)
1150 E. Cheyenne Mntn. Blvd.
Colorado Springs, CO 80906
(719) 540-3820

International Microcircuits
525 Los Coches St.
Milpitas, CA 95035
(408) 263-6300

GE Solid State
724 Route 202
Somerville, NJ 08876
(201) 685-6585

Hughes Aircraft Microelectronics Ctr.
500 Superior Ave.
Newport Beach, CA 92658
(714) 759-2727

IMP, Inc.
2830 N. First St.
San Jose, CA 95134
(408) 434-1362

GigaBit Logic, Inc.
1908 Oak Terrace Lane
Newbury Park, CA 91320
(805) 499-0610

ICI Array Technology, Inc.
1297 Parkmoor Ave.
San Jose, CA 95126
(408) 297-3333

LSI Logic Corp.
1551 McCarthy Blvd.
Milpitas, CA 95035
(408) 433-8000

Gould/AMI
2300 Buckskin Rd.
Pocatello, ID 83201
(208) 233-4690

Integrated Circuit Systems, Inc.
2626 Van Buren Ave.
Valley Forge, PA 19482
(215) 666-1900

Marconi Electronic
45 David Dr.
Hauppauge, NY 11788
(516) 231-7710

Matra Design Semiconductor
2840-100 San Thomas Expwy.
Santa Clara, CA 95051
(408) 986-9000

National Semiconductor Corp.
2900 Semiconductor Dr.
Santa Clara, CA 95052
(408) 721-5884

Pioneer Technologies
9100 Gaither Rd.
Gaithersburg, MD 20877
(301) 921-0660

MCE Semiconductor
1111 Fairfield Dr.
West Palm Beach, FL 33407
(407) 845-2837

NCM Corp.
1500 Wyatt Dr.
Santa Clara, CA 95054
(408) 496-0290

Plessy Semiconductor
1500 Green Hills Rd.
Scotts Valley, CA 95066
(408) 438-2900

Micro Linear Corp.
2092 Concourse Dr.
San Jose, CA 95131
(408) 433-5200

NCR Microelectronics Corp.
2001 Danfield Ct.
Fort Collins, CO 80525
(800) 334-5454

Polycore Electronics
1107 Tourmaline Dr.
Newbury Park, CA 91320
(805) 499-6777

Micro LSI Corp.
2065 Martin Ave., Ste. 101
Santa Clara, CA 95050
(408) 727-7987

Nebula Corp.
33 Lyman St.
Westboro, MA 01581
(508) 366-6558

Raytheon Semiconductor
350 Ellis St.
Mountain View, CA 94039
(415) 968-9211

Micro-Rel
2343 W. 10th Pl.
Tempe, AZ 85281
(602) 968-6411

NEC Electronics, Inc.
401 Ellis St.
Mountain View, CA 94039
(415) 965-6333

Schweber Electronics
CB 1032 Jericho Tpke.
Westbury, NY 11590
(516) 334-7555

Mitsubishi Electronics America
1050 E. Arques Ave.
Sunnyvale, CA 94086
(408) 730-5900

OKI Semiconductor, Inc.
785 North Mary Ave.
Sunnyvale, CA 94086
(408) 720- 1900

Seattle Silicon Corp.
3075 112th Ave. N.E.
Bellevue, WA 98004
(206) 828-4422

Motorola, Inc. ASIC Division
1300 N. Alma School Rd.
Chandler, AZ 85224
(602) 821-4219

Panasonic Industrial Co.
1610 McCandless Dr.
Milpitas, CA 95035
(408) 946-4311

Semiconductor Specialists
195 W. Spangler Ave.
Elmhurst, IL 60126
(312) 279-1000

Sierra Semiconductor
2075 N. Capitol Ave.
San Jose, CA 95132
(408) 263-9300

Standard Microsystems Corp.
35 Marcus Blvd.
Hauppauge, NY 11788
(516) 273-3100

Unicorn Microelectronics
99 Tasman Dr.
San Jose, CA 95134
(408) 433-3388

Signetics Corp./Philips
811 E. Arques Ave.
Sunnyvale, CA 94086
(408) 991-5401

Tachonics Corp.
107 Morgan Lane
Plainsboro, NJ 08536
(609) 275-2509

United Silicon Structures
1971 Concourse Dr.
San Jose, CA 95131
(408) 435-1366

Siliconix, Inc.
2201 Laurelwood Rd.
Santa Clara, CA 95054
(800) 554-5565

Tektronix Integrated Circuits
P.O. Box 14928
Portland, OR 97124
(800) 835-9433

UTMC
1575 Garden of the Gods
Colorado Spgs, CO 80907
(719) 594-8124

Silicon Systems, Inc.
14351 Myford Rd.
Tustin, CA 92680
(714) 731-7110

Texas Instruments, Inc.
P.O. Box 655303
Dallas, TX 75265
(214) 997-2031

Vitesse Semiconductor
741 Calle Plano
Camarillo, CA 93010
(805) 388-3700

Silicon West, Inc.
5470 Anaheim Rd.
Lomg Beach, CA 90815
(213) 494-4588

TLSI, Inc.
790 Park Ave.
Huntington, NY 11743
(516) 549-6300

VLSI Technology, Inc.
1109 McKay Dr.
San Jose, CA 95131
(408) 434-3100

SIS Microelectronics, Inc.
P.O. Box 1432
Longmont, CO 80502
(303) 776-1667

Toshiba America, Inc.
1220 Midas Way
Sunnyvale, CA 94086
(408) 733-3223

VTC, Inc.
2401 E. 86th St.
Bloomington, MN 55425
(612) 851-5200

S-MOS Systems, Inc.
2460 N. First St.
San Jose, CA 95131
(408) 922-0200

TriQuint Semiconductor
P.O. Box 4935 - Group 700
Beaverton, OR 97075
(503) 644-3535

WaferScale Integration
47280 Kato Rd.
Fremont, CA 94538
(415) 656-5400

APPENDIX C

Directory of Programmable Logic Vendors

Actel Corp.
955 E. Arques Ave.
Sunnyvale, CA 94086
(408) 739-1010

Gould/AMI
2300 Buckskin Rd.
Pocatello, ID 93204
(208) 234-6668

Samsung Semiconductor
3725 N. First St.
San Jose, CA 95134
(408) 434-5400

Advanced Micro Devices
901 Thompson Pl.
Sunnyvale, CA 94088
(800) 222-9323

Intel Corp.
1900 Prarie City Rd.
Folsom, CA 95630
(916) 351-6290

SGS-Thompson
1310 Electronics Dr.
Carrollton, TX 75006
(214) 466-7346

Altera Corp.
3525 Monroe St.
Santa Clara, CA 95052
(408) 984-2800

International CMOS Technology
2125 Lundy Ave.
San Jose, CA 95131
(408) 434-0678

Signetics Corp.
811 E. Arques Ave.
Sunnyvale, CA 94086
(408) 991-2000

Atmel Corp.
2125 O'Nel Dr.
San Jose, CA 95131
(408) 441-0311

Lattice Semiconductor Corp.
5555 N.E. Moore Ct.
Hillsboro, OR 97124
(503) 681-0118

Texas Instruments, Inc.
P.O. Box 80966
Dallas, TX 75380
(800) 232-3200

Cypress Semiconductor Corp.
3901 N. First St.
San Jose, CA 95134
(408) 943-2666

National Semiconductor Corp.
2900 Semiconductor Dr.
Santa Clara, CA 95052
(408) 721-5341

Xilinx, Inc.
2069 Hamilton Ave.
San Jose, CA 95125
(408) 559-7778

Exel Microelectronics, Inc.
2150 Commerce Dr.
San Jose, CA 95161
(408) 432-0500

PLX Technology, Inc.
625 Clyde Ave.
Mountain View, CA 94043
(415) 960-0448

Gazelle Microcircuits, Inc.
2300 Owen St.
Santa Clara, CA 95054
(408) 982-0900

Ricoh Corp.
2071 Concourse Dr.
San Jose, CA 95131
(408) 434-6700

APPENDIX D

Directory of CAE Tool Vendors

AB Associates
P.O. Box 82215
Tampa, FL 33682
(813) 932-9853

Aptos Systems
5274 Scotts Valley Dr.
Scotts Valley, CA 95066
(408) 438-2199

Control Data
2800 Old Shakopee Rd.
Minneapolis, MN 55440
(612) 853-3177

Accel Technologies
7358 Trade St.
San Diego, CA 92121
(619) 695-2000

Cadem
1935 N. Buena Vista St.
Burbank, CA 91504
(818)841-9470

Daisy Systems
700 E. Middlefield Rd.
Mountain View, CA 94039
(415) 960-6674

Advanced Microcomputer
2780 S.W. 14th St.
Pompano Beach, FL 33069
(800) 9PC-FREE

Cadence Design Systems
555 River Oaks Pkwy.
San Jose, CA 95134
(800) 283-4080

Data I/O
10525 Willows Rd. N.E.
Redmond, WA 98073
(206) 881-64444

Aida
5155 Old Ironsides Dr.
Santa Clara, CA 95054
(408) 980-5200

CAECO
2945 Oakmead Village Ct.
Santa Clara, CA 95051
(408) 988-0128

ECAD (Cadence)
2455 Augustine Dr.
Santa Clara, CA 95054
(408) 727-0264

Aldec-Automated Logic Design
3525 Old Conejo Rd.
Newbury Park, CA 91320
(805) 499-6867

Calay Systems
16842 Von Karman Ave.
Irvine, CA 92714
(714) 863-1700

Endot
11001 Cedar Ave., Ste 500
Cleveland, OH 44106
(216) 229-8900

Analog Design Tools
1080 East Arques Ave.
Sunnyvale, CA 94086
(800) ANA-LOG4

Case Technology
2141 Landings Dr.
Mountain View, CA 94043
(415) 962-1440

Epic
3080 Olcott St.
Santa Clara, CA 95051
(408) 988-2944

Analogy
P.O. Box 1669
Beaverton, OR 97075
(503) 626-9700

Computervision
100 Crosby Dr.
Bedford, MA 01730
(617) 275-1800

Gateway Design Automation
P.O. Box 573
Westford, MA 01886
(617) 692-9400

186

Genrad
510 Cottonwood Dr.
Milpitas, CA 95035
(408) 432-1000

Mentor Graphics
8500 S.W. Creekside Pl.
Beaverton, OR 97005
(503) 626-7000

Schlumberger CAD/CAM
4251 Plymouth Rd.
Ann Arbor, MI 48106
(313) 995-6000

Hewlett-Packard
1820 Embarcadero Rd.
Palo Alto, CA 94303
(415) 857-1501

Microsim
23175 LaCadena Dr.
Laguna Hills, CA 92653
(714) 770-3022

SDA Systems
555 River Oaks Pkwy.
San Jose, CA 95134
(408) 943-1234

HHB Systems
1000 Wyckoff Ave.
Mahwah, NJ 07450
(201) 848-8000

Omation
1210 E. Campbell Rd.
Richardson, TX 75081
(800) 553-9119

Seattle Silicon Corp.
3075 112th Ave. N.E.
Bellevue, WA 98004
(206) 828-4422

IC Designs
12020 113th Ave. N.E.
Kirkland, WA 98034
(206) 821-9202

OrCAD Systems
1049 S.W. Baseline Rd.
Beaverton, OR 97123
(503) 640-5007

SCSC (Mentor Graphics)
2045 Hamilton Ave.
San Jose, CA 95125
(408) 371-2900

Ikos Systems
145 N. Wolfe Rd.
Sunnyvale, CA 94086
(408) 245-1900

Personal CAD Systems
1290 Parkmoor Ave.
San Jose, CA 95126
(408) 971-1300

Silvar-Lisco
1080 Marsh Rd.
Menlo Park, CA 94025
(415) 324-0700

Intel
3065 Bowers Ave.
Santa Clara, CA 95051
(408) 765-8080

Phase Three Logic
P.O. Box 985
Hillsboro, OR 97123
(503) 640-2422

Tangent Systems
2840 San Thomas Expwy
Santa Clara, CA 95051
(408) 980-0600

Intergraph
One Madison Industrial Park
Huntsville, AL 35807
(205) 772-2000

Racal-Redac
238 Littleton Rd.
Westford, MA 01886
(617) 692-4900

Teradyne Design & Test
321 Harrison Ave.
Boston, MA 02118
(617) 482-2700

Valid Logic Systems
2820 Orchard Pkwy.
San Jose, CA 95134
(408) 432-9400

Viewlogic Systems
275 Boston Post Rd. W.
Marlboro, MA 01752
(617) 480-0881

APPENDIX E

Directory of Package and Assembly Vendors

Addison Engineering, Inc.
656 E. Taylor Ave.
Sunnyvale, CA 94086
(408) 749-1000

Nepenthe
2471 E. Bayshore Rd.
Palo Alto, CA 94303
(415) 856-9332

Textool/3M
1001 Fountain Pkwy.
Grand Prarie, TX 75050
(214) 647-0392

Dupont Electronics
Box 80013
Wilmington, DE 19880
(800) 237-4357

NTK
349 Cobalt Way, Ste. 304
Sunnyvale, CA 94086
(408) 736-7205

Dyne-Sem
1875 S. Grant St., Ste. 630
San Mateo, CA 94402
(415) 574-4477

Pacific Hybrid
10575 S.W. Cascade Blvd.
Portland, OR 97223
(503) 684-5657

Ibidin USA Corp.
2727 Walsh Ave., Ste. 203
Santa Clara, CA 95051
(408) 748-7755

Pantronix Corp.
145 Rio Robles Dr.
San Jose, CA 95134
(408) 263-1711

Indy Electronics
400 Industrial Park Dr.
Manteca, CA 95336
(209) 239-4444

Seiko-Epson (S-MOS)
2460 N. First St.
San Jose, CA 95131
(408) 922-0200

Jade (Kidde)
1120 Industrial Hwy.
Southampton, PA 18966
(215) 322-9020

Shinko Electric America, Inc.
4701 Patrick Henry Dr.
Santa Clara, CA 95054
(408) 727-2133

Kyocera America, Inc.
8611 Balboa
San Diego, CA 92123
(619) 576-2600

Swire Technologies
4800 Great America Pkwy.
Santa Clara, CA 95054
(408) 748-9933

APPENDIX F

Glossary of ASIC Terms and Acronyms

ASIC An acronym for application specific integrated circuit. ASICs may take the form of custom, semicustom, or programmable ICs.

Asynchronous Any signal lacking a regular, predictable timing relationship. The output state of a circuit may be independent of the clock signal, and the operational speed of a signal depends only on its propagation delay through a network, rather than on clock pulses.

ATPG An acronym for automatic test pattern generation. ATPG is generally used to augment functional test patterns to increase the level of fault coverage.

Back-annotation The automatic process of updating simulation files with delays extracted from the actual post-layout wiring capacitance.

Behavioral description A model of a device or function in terms of algorithms or mathematical equations.

Benchmark A standard test by which products are evaluated and compared.

BiCMOS A mixed-technology process generally used to fabricate ASICs consisting of bipolar I/O cells and a CMOS core. The bipolar I/O transistors provide fast switching and high output drive, while the CMOS transistors comprising the core cells provide high functional density and also conserve power.

Bipolar A process technology employing two-junction transistors (npn, pnp).

Bonding diagram A document specifying the I/O pad-to-package post connectivity (defines the package pin-out for the device).

Boundary scan A testability enhancement technique whereby the device incorporates a scan path through the device inputs and outputs. The technique is used primarily for facilitating board-level testing.

Bounding box A polygon defining the borders or outline of a cell.

Buffer A low-impedance inverting driver circuit that can supply substantially more output current than the basic circuit. The buffer element is used for driving heavily loaded circuits or minimizing rise-time deterioration due to capacitive loading.

CAE An acronym for computer-aided engineering.

Cell A predefined layout of circuit elements that implements a specific electrical function.

Cell library A collection of cells whose characteristics are generally specific to an ASIC vendor.

CFI An acronym for the CAD Framework Initiative; a consortium of industry participants whose charter is to define standards for the interfacing of design tools.

Chip carrier A surface-mounted package technology typified by leads on all four sides of the package body.

Clock skew The phase shift in a single clock distribution network resulting from the different delays in clock driving elements and/or different distribution paths.

Clock tree A clock distribution technique that minimizes clock skew.

CMOS An acronym for complementary metal oxide semiconductor; a circuit technology offering low quiescent power dissipation.

Core The active (or used) area of an ASIC or the area excluding the I/O pad ring.

Core-limited Core-limited designs result when the active area of the chip determines the minimum die size (contrast with *pad-limited*).

191

Critical path The longest path in a circuit network. The critical path propagation delay limits the maximum clock frequency for the device.

CSIC An acronym for customer specific integrated circuit (an alternate term for ASIC).

Current density The current-carrying capacity of a wire (metal interconnect on an ASIC). Should the current exceed the wire's capacity, metallurgical failures can result.

Cut and go A prototyping method where dice are packaged and shipped to a customer prior to device testing.

Data path A bus-oriented circuit architecture optimized for pipelined or bit-parallel operations. Data paths typically include data processing elements such as ALUs, muxes, shift registers and other register-transfer functions.

Derating factor A factor used to derate nominal or typical performance parameters. A typical derating factor for the combination of worst-case temperature ($+125°C$), voltage (5.5 V), and process corners is 2.5X the nominal delays for a given cell or path.

Defect A fault condition resulting from contamination or a manufacturing anomaly that may or may not affect circuit operation.

Defect density The number of defects statistically present in a given area of the wafer. A typical defect density for a process in a class 10 or better clean room is one to two defects per square centimeter of wafer area.

DESC An acronym for the Defense Electronics Supply Center.

Design rules The recipe for a process. Design rules specify minimum width and spacing requirements for the polygons comprising the physical layout.

Die Also called a chip, a die is an individual circuit sawn or broken from a wafer that contains an array of such circuits or devices.

Die attach The process of attaching the die to the package cavity or substrate. Die attach materials include epoxy, gold, and silver-filled glass.

DRC An acronym for design rule checker; a program that performs process design rule spacing checks on the completed circuit layout to ensure that the design artwork will be manufacturable in the specified process.

Dynamic 1. Any type of device testing during which the clock is applied. 2. A memory element in which logic state storage depends on capacitively charged circuit elements. These elements must be continually refreshed or recharged at regular intervals.

E-beam Short for electron beam; an evaporation technique that uses the energy of a focused electron beam to provide the required heat.

ECL An acronym for emitter-coupled logic; a bipolar circuit technique in which the emitters of the input logic transistors are coupled to the emitter of a reference transistor. ECL circuits are characterized by extremely fast switching rates due to low voltage swing requirements, but because the transistors are always partially on, they consume a large amount of power.

EDA An acronym for electronic design automation; refers to computer-aided engineering tools.

EDIF An acronym for electronic data interchange format; a standard, neutral format for transporting data created by one design tool to another.

ERC An acronym for electrical rules checker; a program that checks a circuit layout for electrical rules violations such as excessive fan-out, opens, and shorts.

ESD An acronym for electrostatic discharge. When electrostatic-induced voltages are discharged at a device's input pins, physical damage to the device is likely to result. A condition also known as a "zap."

Fault A manufacturing defect that can have the effect of causing an open or a short in a circuit, resulting in functional failure.

Fault coverage Refers to the percentage of possible fault conditions that can be detected by a production test program.

Fault grading A computer-automated process of determining fault coverage by simulating faults that are modeled as circuit nodes which are *stuck* at a logic 0 (open) or 1 (short).

Feedthrough A region in a circuit layout dedicated for crossing a cell with a signal such that the line is electrically isolated from the cell.

Floating node A gate input or output that is erroneously left unconnected, resulting in functional failure. Floating nodes will generally *float* to a logic high state. ERC programs catch such layout errors, though they may also be caused by incomplete contacts in the processing.

Floor planning The process of placing functional blocks within the chip layout area and allocating interconnect routing between them such that an optimum layout is achieved.

Foundry A semiconductor fabrication facility that makes its excess manufacturing capacity available to other semiconductor companies or customers with their own process-compatible tooling.

FPGA An acronym for field programmable gate array; an electrically user-programmable device.

Frameworks A CAE environment that provides a common design database and user interface, allowing the transfer of design data from one tool to the next without the risk of translation errors.

Full-custom An integrated circuit design in which each circuit element is individually *drawn* and positioned in the chip layout. All mask layers are specific to the design.

GaAs A nonsilicon process technology characterized by ultrafast operating frequencies (gigaHertz).

Gate A circuit having two or more inputs and a single output, the output state being a function of the combination of logic signals at the inputs. The fundamental logic gate types perform Boolean functions such as AND, OR, NAND, and NOR.

Gate array A semicustom ASIC technology that utilizes prefabricated wafers processed up to the final metal interconnect layers. These generic *master slices* are then personalized (customized) as a function of the user's design database, which defines the connectivity of the array of prefabricated transistors (or gates, depending on the array architecture).

Gate equivalency The number of gates a design implements or utilizes on an ASIC or ASIC cell.

GDSII A design data format used to generate artwork for mask making.

Glitch An input transition or voltage spike that occurs in a time period that is shorter than the delay through the affected logic element. Glitches can propagate to primary outputs, causing functional failure.

Golden simulator The simulator by which a vendor's cell library performance is characterized, qualified, and guaranteed.

Ground bounce A switching transient caused by a large number of high-drive IC outputs switching simultaneously.

Guardband The margin applied to device specifications in consideration of instrumentation precision in electrical testing.

HDL An acronym for hardware description language; a high-level behavioral abstract of a design, defined in a software algorithm. Contrast with *structural description*.

Hierarchical description A structural design description consisting of multiple nested levels of logic. For example, a microprocessor block contains a program counter block that consists of flip-flop blocks, which consist of gate blocks.

ILB An acronym for inner-lead bonding; a process in the TAB manufacturing sequence that connects the tape's inner leads to the chip.

Instance An occurrence of a cell in a schematic, netlist, or layout.

I/O An abbreviation for Input and Output signals.

JEDEC A standards organization whose charter is to eliminate misunderstandings between manufacturers and purchasers.

JIT An acronym for just-in-time delivery; an inventory management technique in which product is delivered exactly when it is needed on the production floor. Requires close cooperation with the vendor in areas of material requirements forecasting and quality assurance.

JTAG An acronym for the Joint Test Action Group; authors of the boundary scan standard. See *boundary scan*.

Latch-up A condition in which the voltage in a circuit does not return to the supply voltage when a transistor is switched from saturation to cut off, resulting in functional failure and possible circuit burnout.

Load The resistance and/or capacitance that the inputs of one or more devices present to the output of a driving device to which it is connected.

LPE An acronym for layout parameter extraction; a program that extracts the parasitic resistance and capacitance of a node or net and outputs a SPICE netlist for use in a detailed analysis of critical path performance.

LSSD An acronym for level-sensitive scan design; IBM's variant of the scan test technique.

LVS An acronym for layout versus schematic; a program that performs a comparison between the schematic netlist and its corresponding physical design database (layout) and reports any discrepancies.

Macro cell/hard macro Also referred to as a core function, a macro is a complex ASIC cell performing a function originally offered as a standard catalog component. *Hard* macros are so called because their physical layout is fixed in the design rules for which they were originally designed and characterized.

Macro function/soft macro Unlike hard macros, which are defined at the physical layout level, *soft* macros are defined at the cell library and netlist level. A soft macro may implement the same electrical function as that of a hard macro, but has no predetermined physical layout.

Maintenance A fee charged by software vendors to provide technical support, bug fixes, and updates for their tools.

Mask A glass plate template consisting of clear or opaque areas that respectively allow or prevent light to shine through. The masks are aligned with existing patterns on the wafer and are used to expose photoresist for the defining of circuit elements and their connectivity.

MCM An acronym for multichip module; typically, a ceramic substrate that carries several die.

Metallization The process of depositing layers of high-conductivity material (aluminum-silicon) used to interconnect circuit elements or cells on a chip.

Mixed mode A simulation system that combines both transistor-level and gate-level models in the same analysis.

Mixed signal A design that integrates both digital and analog circuits on the same device, or in the same analysis (see *mixed mode*).

Moat The area in a chip layout that is reserved for the routing of I/O signals from the core to the I/O pads.

Model   A behavioral representation of a physical device (including its timing characteristics).

Module   1. A cell or function. 2. A packaging technology that combines several individually packaged devices into a single subassembly.

Module generator   A software tool that automatically configures the physical layout of a cell in accordance with a specific set of design parameters. For example, a flip-flop may be configured with or without preset and clear lines; a RAM cell may be generated in the desired organization.

NDA   An acronym for nondisclosure agreement; a legal document safeguarding proprietary interests in design or other intellectual data or property.

Net   Short for network; a circuit path.

Netlist   An ASCII listing of a design's constituent cells and their connectivity.

Node   A terminal of a circuit element or any branch of a net.

NRE   An acronym for nonrecurring engineering; refers to the activity and associated cost of a development effort, such as that for the development of ASIC prototypes.

OLB   An acronym for outer-lead bond; a process in the TAB manufacturing sequence that connects the tape's outer leads to the printed circuit board or substrate.

Pad   A metallized area in the periphery of the chip that is used to connect the I/O circuitry to the package or substrate.

Pad-limited   Pad-limited designs result when the number of input, output, power, and ground pads in the periphery of the chip determine the minimum die size. Contrast with *core-limited*.

Pad ring   The area around the perimeter of the chip comprising the pads.

Parasitic capacitance The capacitance in a net attributable to the capacitance of the metal interconnect or routing.

Partitioning The process of allocating circuit sections and subsections to specific ASIC implementations.

Passivation A layer of material (typically glass) deposited over a completed IC to stabilize its surface and provide protection from contamination.

PCB An acronym for printed circuit board.

PCM An acronym for process control monitor; a test structure on the wafer that is used to verify that all process parameters are within controlled limits.

Personalization
The customization of a generic gate array device, which is accomplished via metal interconnect layers.

PGA An acronym for pin-grid array; a through-hole mounted package technology in which the leads of the package (arranged in a grid) extend downward from the underside of the package body.

Physical design The physical implementation of an integrated circuit layout in terms of the geometric elements comprising transistors, cells, and blocks, as well as their placement and routing.

Placement The physical location of a cell or block in a chip layout.

Platform Refers to the type of computer hardware or workstation used to run the applications software or design tools.

PLCC
An acronym for plastic leaded chip carrier; a common low-cost, surface-mounted package technology.

Polycrystalline silicon Silicon composed of many single crystals having a random arrangement. Also known as polysilicon, or simply poly. Polysilicon is often used as an intracell interconnect material and also performs well as a capacitor plate.

PQFP An acronym for plastic quad flat pack; a common low-cost, plastic, surface-mounted package technology that facilitates high pin counts.

Primary output An output signal that is accessed directly from a package pin.

Primitive A low-level function such as a gate.

Probe card A chip-contacting test fixture that enables the testing of die while they are still on the wafer.

Prototype An original design or first operating model intended for evaluation of its form, fit, and function for a particular application. Prototype ASIC devices are representative of the final production units.

Race condition The condition that results when a signal is propagated through two or more logic or memory elements in the same clock period, thus violating the timing requirements for proper circuit operation. Also called a timing hazard.

Routing The interconnecting paths between cells.

RTL An acronym for register-transfer level; a design description that combines a behavioral description with a structural description. An RTL description describes the data flow from register to register.

Scan test A design-for-testability technique whereby test patterns can be shifted into sequential circuit elements (flip-flops), clocked and then shifted out for comparison with the expected results.

Scribe line The area on a wafer that separates adjacent die locations. The scribe line, also called a street, is the line along which the wafer is scribed or sawn for yielding individual die.

Sea of gates A gate array architecture that features a continuous array of transistors. Sea-of-gates arrays utilize unused transistor sites for routing, as they have no dedicated routing channels.

Semicustom Refers to any ASIC methodology that utilizes prefabricated and characterized circuit elements, thus requiring only element interconnect for customization.

SEU An acronym for single-event upset; a radiation-induced condition that causes a bistable circuit element (flip-flop or memory bit) to change its state.

Silicon compiler A design tool that, when provided with a high-level design description, compiles and/or synthesizes all the necessary design views, including those for physical design and simulation. See *module generator*.

Simulation or test vector A stimulus pattern that, when applied to a circuit's inputs and is operated upon, produces an output result, which is compared against the expected response.

SOI An acronym for silicon on insulator; a CMOS process technology in which transistors are formed in the *grown* layer of silicon on top of the substrate material.

SOS An acronym for silicon on sapphire; a CMOS process technology in which transistors are formed in the *grown* layer of silicon on top of the substrate material.

SOW An acronym for statement of work; a document specifying the tasks to be completed by both the ASIC vendor and the designer.

SPC An acronym for statistical process control; a product quality enhancement methodology whereby all in-line manufacturing processes are monitored and controlled for conformance to specified requirements, thus reducing process variability and improving predictability.

SPICE An acronym for Simulation Program with Integrated Circuit Emphasis; a circuit simulator.

Standard cell A primitive functional element such as a gate or latch that is characterized by fixed physical and electrical characteristics.

Static An unclocked mode or condition.

Step coverage Refers to the quality of the interconnect metallization layers as they cover the vertical steps that occur when they go into and out of contact holes and over other metal lines (metal lines tend to thin out at the knee of such steps, making them susceptible to open circuit conditions).

Structural description The intermediate design description level falling between behavioral- and gate-level descriptions. The structural level describes the data flow from register to register and distinguishes the data paths from the control paths.

Surface mount A device package-to-PCB mounting technology.

Symbol A graphical representation of a cell featuring its bounding box and I/O ports. Symbols are used primarily in schematic editing.

Synchronous A design quality whereby the performance of operations is controlled by regular clock intervals.

Synthesis The translation of a high-level design description (consisting of state transition machines, truth tables, and/or Boolean equations) into a specific gate-level logic implementation.

TAB An acronym for tape-automated bonding; an integrated circuit packaging technology that features individual die mounted in a roll of continuous sprocketed film, from which the die is excised and mounted onto a PCB or other substrate.

Turnkey design A service provided by most ASIC vendors where they design and produce an ASIC device based on design specifications provided by the customer.

VHDL An acronym for VHSIC Hardware Description Language; a standard, technology-independent design description language used to generate a specification that can ideally be targeted to any number of ASIC vendors who could produce a functionally equivalent chip, according to that language-based specification. See *HDL*.

Workstation See *platform*.

Yield The number of units surviving screening operations.

APPENDIX G

Read More About It

The following periodicals are excellent sources for information on new developments and offerings in ASIC and related technologies. Most publishers offer free subscriptions to those who satisfy their qualification criteria.

ASIC Technology & News
480 San Antonio Road, Suite 245
Mountain View, CA 94040
(415) 949-2742

Electronics
VNU Business Publications
Ten Holland Drive
Hasbrouck Heights, NJ 07064
(201) 393-6000

Electronic Engineering Times
CMP Publications, Inc.
600 Community Drive
Manhasset, NY 11030
(506) 562-5000

Computer Design
PennWell Publishing Company
One Technology Park Drive
Westford, MA 01886
(508) 692-0700

Electronic Business
Cahners Publishing Company
275 Washington Street
Newton, MA 02158-1630
(617) 964-3030

Electronic News
Chilton Co.
825 7th Avenue
New York, NY 10019
(215) 630-0951

Design Automation
Miller Freeman Publications
600 Harrison Street
San Francisco, CA 94107
(415) 905-2200

Electronic Buyers' News
CMP Publications, Inc.
600 Community Drive
Manhasset, NY 11030
(506) 562-5000

Electronic Packaging
Cahners Publishing Company
275 Washington Street
Newton, MA 02158-1630
(617) 964-3030

Digital Design
Digital Design Publishing Corp.
1050 Commonwealth Avenue
Boston, MA 02215
(800) 223-7126

Electronic Design
Penton Publishing Company
1100 Superior Avenue
Cleveland, OH 44114-2543
(216) 696-7000

Electronic Products
Hearst Business Communications
645 Stewart Avenue
Garden City, NY 11530
(516) 227-1300

EDN
The Cahners Publishing Company
275 Washington Street
Newton, MA 02158-1630
(617) 964-3030

Electronic Engineering Manager
CMP Publications, Inc.
600 Community Drive
Manhasset, NY 11030
(506) 562-5000

Electronics Purchasing
Cahners Publishing Company
275 Washington Street
Newton, MA 02158-1630
(617) 964-3030

Electronics Test
Miller Freeman Publications
600 Harrison Street
San Francisco, CA 94107
(415) 905-2200

IEEE Spectrum
The Institute of Electrical and Electronics Engineers, Inc.
345 East 47th Street
New York, NY 10017
(212) 705-7555

Military & Aerospace Electronics
Sentry Publishing Company, Inc.
1900 West Park Drive
Westborough, MA 01581
(508) 366-2031

Printed Circuit Design
Miller Freeman Publications
600 Harrison Street
San Francisco, CA 94107
(415) 905-2200

Solid State Technology
PennWell Publishing Company
One Technology Park Drive
Westford, MA 01886
(508) 692-0700